MILITARY
JET AIRCRAFT

MILITARY JET AIRCRAFT

1945 TO THE PRESENT DAY

ROBERT JACKSON

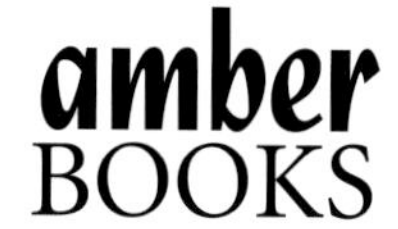

This revised Amber edition first published in 2019

Published by Amber Books Ltd
United House
North Road
London N7 9DP
United Kingdom
www.amberbooks.co.uk
Instagram: amberbooksltd
Facebook: www.facebook.com/amberbooks
Twitter: @amberbooks

First published in 2003 as *Military Jets: Design and Development*

ISBN: 978-1-78274-882-3

Project Editors: James Bennett and Michael Spilling
Designer: Colin Hawes

Printed in China

PICTURE CREDITS:
All photographs and artworks Amber Books Ltd/Art-Tech except:

Department of Defense: 82 (bottom), 104 & 105, 108 & 109.
Cody Images: 7, 9, 10/11, 22, 26, 30, 34, 47, 54 (top), 59, 60, 64, 68, 72, 79, 86, 92, 93, 97, 100, 102, 104, 140, 146, 151, 157 (bottom), 161, 164 (bottom), 169, 172, 181, 185, 188, 191, 196, 211, 213, 230, 236, 244, 246, 250.
Military Vizualisations, Inc: 102/103, 106/107.

Contents

Introduction	6
United States of America	10
United Kingdom	110
France	142
International	172
Sweden	198
USSR/Russia	214
Index	254

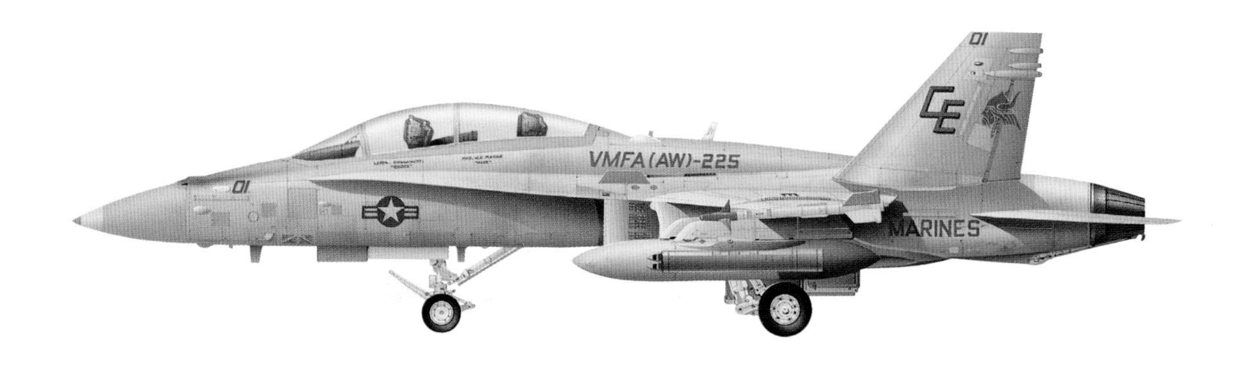

Introduction

Although jet engine technology was pioneered in Britain by Frank Whittle, it was the Germans who made the first successful marriage of a turbojet with an airframe.

Developed as a private venture in parallel with the rocket-powered He 176, the Heinkel He 178 was only ever intended as an experimental test-bed, although it made its mark in the history books when, on 27 August 1939, Flugkapitan Erich Warsitz took off in the world's first jet-powered aircraft and circled the Heinkel factory at Rostock-Marienehe before landing safely. However, although the He 178 was demonstrated before leading officials of the Germany Air Ministry in October, little official interest was shown, and there were technical problems associated with the fuselage-mounted turbojet. Development of the He 178 was therefore abandoned in favour of a project with twin wing-mounted jet engines, the Heinkel He 280, which made its first turbojet-powered flight on 2 April 1941 – six weeks before the maiden flight of Britain's first jet aircraft, the Gloster/Whittle E28/39.

The development of the new turbojet technology in Britain and the United States during World War II was slow, mainly because Allied aircraft industries were fully occupied with the rapid development of new combat types that depended on existing piston-engine technology. By 1943, however, German development of jet aircraft began to gather momentum for two reasons. The first was the need for a new fighter with a performance that would enable it to break through the increasingly powerful Allied fighter escort screen before engaging the formations of American daylight

Left: **The Messerschmitt Me 262, the world's first turbojet-powered fighter to see combat, was a very advanced aerodynamic design for its day. It was plagued by its Jumo 004 engines, which had a life of only 25 hours. The leading aircraft in this photograph is a two-seat Me 262B-1a.**

bombers, and the second was a pressing requirement for a fast bomber/reconnaissance aircraft whose speed and ceiling would render it virtually immune from interception. This requirement became even more urgent early in 1944, when the remnants of the Luftwaffe's bomber forces were decimated in the so-called 'Little Blitz' against England, and when reconnaissance of objectives in the British Isles in the months prior to D-Day was almost totally denied by effective air defences. These requirements crystallized, on the one hand, in the Messerschmitt Me 262 fighter/attack aircraft, which was selected for production in preference to the He 280, and on the other in the Arado Ar 234 bomber/reconnaissance aircraft. Both aircraft entered service on an experimental basis in the summer of 1944, as did Britain's first jet fighter, the Gloster Meteor.

In the years immediately after World War II, the power and the menace of the strategic bomber led to the rapid development of the jet fighter, which, thanks to advances in metallurgy, now had engines that were infinitely more reliable and powerful than those developed during the war.

The jet age saw the rebirth of the pure interceptor, aircraft such as America's F-86 Sabre, the Soviet Union's MiG-15 and Britain's Hawker Hunter, high speed gun platforms whose sole purpose, originally, was to climb fast enough and high enough to destroy the strategic bomber. But such a task had to be undertaken by day and night, in all weathers, and this in turn led to the evolution of the 'weapon system', a fully integrated combination of airframe, engine, weapons, fire control systems and avionics. Early examples of the weapon system were the Northrop F-89 Scorpion, the Lockheed F-94 Starfire and the Gloster Javelin.

In the mid 1950s yet another concept of air warfare came about, partly as a result of the lessons learned during the Korean War, and partly because of the wildly escalating cost of developing new combat aircraft. This reached fruition in such highly successful types as the McDonnell F-4 Phantom and the Dassault Mirage family, whose basic airframe/engine combination was designed from the outset to support long-term development compatible with a wide variety of operational requirements. The high cost of developing new and complex airborne weapon systems also brought about international co-operation on an unprecedented scale, with a highly beneficial pooling of brains, expertise and financial resources which was to result in the production of advanced and versatile military aircraft such as the Panavia Tornado and the Eurofighter Typhoon.

SUCCESSES AND FAILURES

The list of success stories in the field of military jet aircraft production since 1945 is a long one. In terms of longevity Britain's Canberra and America's B-52 are unmatched, both remaining in first-line service more than half a century after their prototypes first flew, while for sheer proliferation Russia's MiG-15 remains unsurpassed, having been produced in far greater quantities than any other combat type. To this list should be added the postwar export successes of France's military aircraft industry, whose products have admirably upheld the French tradition for building machines which combine a high degree of potency with aesthetic appeal.

Against the successes, however, must be measured the many failures, some resulting from changing policies, others from prohibitive costs, and still others from political misconceptions – or a combination of all three. Some of the greatest political misconceptions of all time were contained in the British Defence White Paper of 1957, which announced the phasing out of manned aircraft in favour of missiles and sounded the death knell for several promising projects which might have become Britain's military best-sellers in the 1960s. None of them, however, had such a profound and damaging effect on the British aerospace industry as the later cancellation of the superb TSR-2, leaving a gap which was not

Left: **Designed to replace the venerable A-7 Corsair II in US Navy and Marine Corps service, the F/A-18 Hornet is a combat-proven design and has been the subject of substantial export orders. Customers include Canada, Finland, Kuwait and Switzerland.**

filled, and then not entirely, until the advent of the Tornado nearly 20 years later. The Americans had their troubles too, as did the Soviets, but the budgets of both these powers were far better placed to support the cancellation of advanced aircraft projects than those of economically weaker nations. The Americans almost fell into the 'all-missile' trap when they cancelled the variable-geometry Rockwell B-1 in favour of cruise missiles, until a successor administration saw the light and resurrected the project as the B-1B. In many respects the Russians seem to have had a shrewder appreciation of defence requirements. Lacking aircraft carriers, they developed the Tupolev Tu-22M Backfire supersonic bomber, which with flight refuelling placed targets thousand of miles from the Russian homeland within their offensive reach.

FULL CIRCLE

With the deployment of aircraft like the Backfire, specialization returned to the world of military aircraft design. During World War II, the Germans designed the heavily-armoured Henschel Hs 129 specifically to kill tanks; its modern counterpart is the Fairchild Republic A-10 Thunderbolt II. In 1940, the Spitfire, Hurricane and Bf 109 were all what would now be clearly defined as air superiority fighters; in the 1970s this category returned with the development of such types as the McDonnell

Above: **At the cutting edge of aviation technology, the F-22 Raptor is without doubt the most advanced combat aircraft in the world. Congress questioned the wisdom of developing such a hugely expensive aircraft at a time of reduced high-technology threats to the USA.**

Douglas F-15 Eagle and Mikoyan MiG-29. The wheel had turned full circle.

The story of military aircraft design since 1945 has been one of enormous technological progress, reaching new heights with aircraft like the V/STOL Harrier, the F-117 stealth aircraft and the F-22 Raptor, the latter combining every aspect of aviation technology devised so far. But it has also been a story of compromise, with defence economies dictating the need to adapt and upgrade existing designs to fill the gap created by the cancellation of more advanced projects, usually for reasons of soaring cost.

It is already clear where military jet aircraft design is heading in the 21st century. Already, reconnaissance aircraft are in operation which are controlled entirely from the ground, and it will not be long before similar remotely-piloted attack aircraft are also in use. Just as, in World War I, fighters were developed to destroy reconnaissance machines, remotely-controlled interceptors will be developed to destroy remotely-controlled reconnaissance drones. War between robots is no longer science fiction. It is here; it is now.

United States of America

In the early post-war years in the United States, from 1946, research and development facilities were concentrated at Muroc, a vast dry lake in California.

From 1946 to 1958, by which time its name had changed to Edwards Air Force Base, Muroc played a major role in the supersonic breakthrough that followed hard on the heels of the turbojet revolution. The act of centralizing aerodynamic research and test flying at Muroc made possible the evolution of a series of operational combat aircraft which, over the next two decades, were to put the United States a long way in front of any other major aircraft-producing nation. Research at Muroc/Edwards was incorporated in the generation of Mach 2 plus combat aircraft that were to equip the first-line squadrons of the US Navy and USAF by 1960, establishing a lead that the US would never lose, or even come close to losing, until the advent of advanced Soviet air superiority fighters in the 1980s.

Left: **A Rockwell B-1B Lancer climbs away after a fast run over a desert bombing range. The B-1B project was resurrected by President Ronald Reagan after having been cancelled by a previous administration, and became a very viable weapons system.**

Lockheed F-80 Shooting Star

The F-80 pilot sat on a simple ejection seat, within a lightly pressurized cockpit. Without suitable air conditioning or ventilation, the cockpit soon became overheated in hot and humid outside conditions.

The forward fuselage of the F-80C contained a concentrated package of six 12.7mm (0.50in) guns, with 300 rounds per gun. The underwing weapons stations could carry up to ten 12.7cm (5in) high velocity aircraft rockets (HVARs). These were ideally suited to ground attack tasks, and particularly effective against rolling stock.

A characteristic of the Shooting Star family, wing tanks were a necessity during operations over Korea. Each 'Misawa' tank contained 757 litres (200 US gallons) and replaced the standard 'teardrop' tank, which had contained only 625 litres (165 US gallons). The gaudy red stripes on these tanks, and on the tail, identify this F-80C as an aircraft of the 36th Fighter Bomber Squadron, 8th Fighter Bomber Wing.

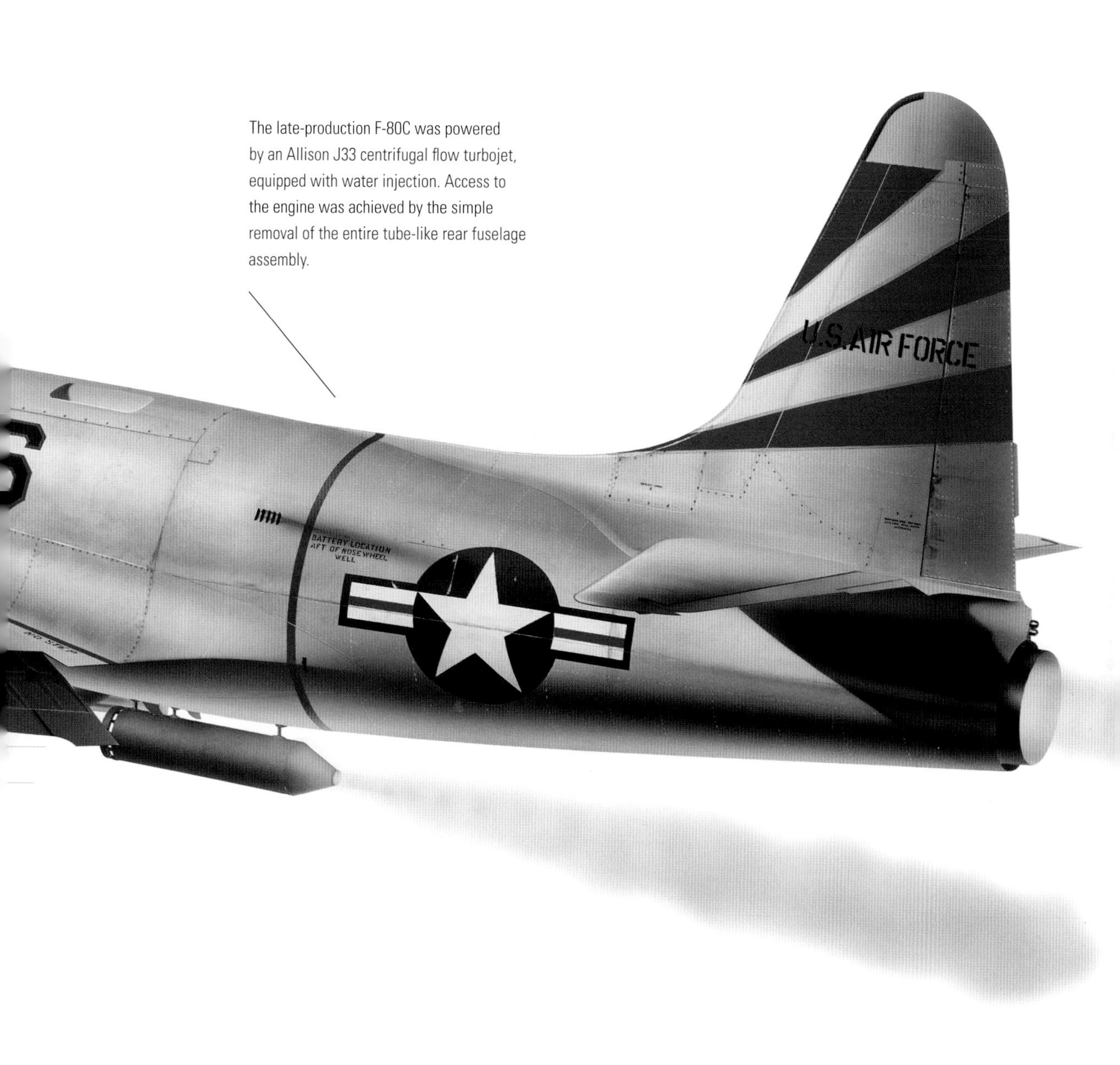

The late-production F-80C was powered by an Allison J33 centrifugal flow turbojet, equipped with water injection. Access to the engine was achieved by the simple removal of the entire tube-like rear fuselage assembly.

Above: **The Lockheed F-80 was America's first operational jet fighter. Pictured here is the F-80B version, which was the first to feature an ejection seat. The F-80B was quickly supplanted by the F-80C, which became the major production version.**

America's first fully operational jet fighter was the Lockheed P-80 Shooting Star, which like its British counterparts was of very conventional design and was to become the workhorse of the American tactical fighter-bomber and fighter interceptor squadrons for five years after World War II. The prototype XP-80 was designed around a de Havilland H-1 turbojet, which was supplied to the United States in July 1943, and the aircraft was completed in just 143 days, making its first flight on 9 January 1944. In April 1945 two YP-80s were sent to England, where they were attached to the Eighth Air Force, and two more went to Italy, but none experienced any operational flying in Europe before the war's end.

Early production P-80As entered USAAF service late in 1945 with the 412th Fighter Group, which became the 1st Fighter Group in July 1946 and comprised the 27th, 71st and 94th Fighter Squadrons. On 12 July 1948, as part of a reinforcement plan following the Soviet blockade of Berlin, 16 Lockheed F-80A Shooting Stars of the 56th Fighter Wing (Lt Col David Schilling), a unit then assigned to Strategic Air Command, left Selfridge Field,

Michigan, and flew via Dow AFB, Maine, Goose Bay in Labrador, Bluie West One in Greenland and Reykjavik in Iceland to make their UK landfall at Stornoway after a total transatlantic flight time of 5 hours 15 minutes. On 21 July the fighters flew on to Odiham, Hampshire. After their Odiham visit the F-80s flew on to Fürstenfeldbruck, where they spent six weeks on exercises, including fighter affiliation with B-29s, before returning to the USA. Early in August 72 F-80s of the 36th Fighter Wing were shipped to Glasgow in the aircraft carrier USS *Sicily* and the US Army transport vessel *Kirschbaum*, and after being offloaded and overhauled they departed for Fürstenfeldbruck between 13 and 20 August.

The P-80A was followed by the P-80B; the major production version was the F-80C (the P for 'pursuit' prefix having changed to the much more logical F for 'fighter' in the meantime). The F-80C was the fighter-bomber workhorse of the Korean War, flying 15,000 sorties in the first four months alone. Pilots found the aircraft ideal for strafing, but the F-80 was rarely able to out-manoeuvre North Korea's piston-engined Yakovlev and Lavochkin fighters, nor were the American jets initially equipped to carry bombs or rockets for effective ground attack work. In spite of these deficiencies, F-80s managed to claim a number of North Korean aircraft, the first on the third day of the war, 28 June 1950, when the 35th Fighter Squadron, nicknamed the 'Panthers' and operating out of Itazuke in Japan, became the first American jet squadron to destroy an enemy aircraft. The engagement took place while the F-80s were protecting a flight of North American Twin Mustangs. Captain Raymond E. Schillereff led four aircraft into the Seoul area and caught a quartet of Ilyushin Il-10s which were attempting to interfere with US transport aircraft embarking civilians at Seoul's Kimpo airfield; all four Il-10s were shot down. The Shooting Star was assured of its place in history when 1st Lt Russell Brown of the 51st Fighter Wing shot down a MiG-15 jet fighter on 8 November 1950, during history's first jet-versus-jet battle. Other sporadic attacks, against which the Shooting Stars were the main USAF defence, were to follow during the early part of the war, until the arrival of the more capable F-86A Sabre in December 1950.

Above: **This RF-80A was flown by the 15th Tactical Reconnaissance Squadron, which was based at Kimpo, Korea, in 1952. The Shooting Star performed excellent service in Korea, being eventually replaced in the ground attack role by the F-84 Thunderjet.**

The RF-80C was a photo-reconnaissance version. Total production of the Shooting Star was 1718, many being later converted to target drones.

Specification: Lockheed F-80C Shooting Star	
Type:	single-seat fighter-bomber
Powerplant:	one 2449kg (5400lb) thrust Allison J33-A-35 turbojet
Performance:	maximum speed 956km/h (594mph) at sea level; service ceiling 14,265m (46,800ft); range: 1328km (825 miles)
Weights:	empty 3819kg (8420lb); maximum take-off 7644kg (16,856lb)
Dimensions:	wing span 11.81m (38ft 9in); length 10.49m (34ft 5in); height 3.43m (11ft 3in); wing area 22.07m² (237.6 sq ft)
Armament:	Six 12.7mm (0.50in) machine guns, plus two 454kg (1000lb) bombs and eight rockets

Republic F-84 Thunderjet

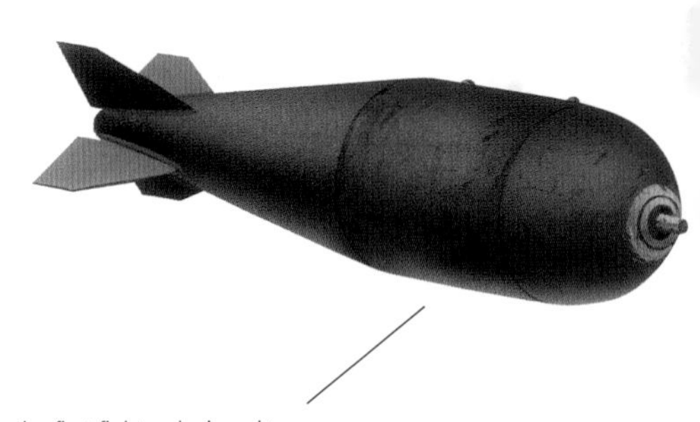

The F-84G was the first fighter designed to deliver a nuclear weapon and, although it was never called upon to do so in anger, it retained an impressive conventional capability, which it exercised to devastating effect in Korea.

The F-84G pictured here, 51-1111 'Five Aces' (so named because of its unusual serial number) served with the 69th Fighter Bomber Squadron, 58th Fighter Bomber Wing, at Taegu in Korea, where it was flown by Lt Jim Simpson on many of his 56 operational missions.

The cockpit of an F-84G was an enviable position for a young pilot. Although denied the glamour of the sleek Sabre, an F-84G could be an intoxicating aircraft to fly, with excellent stability even at high speed at low level. The only problem was its take-off run, which with a fully laden aircraft was long and sometimes marginal. The problem was alleviated by rocket-assisted take-off.

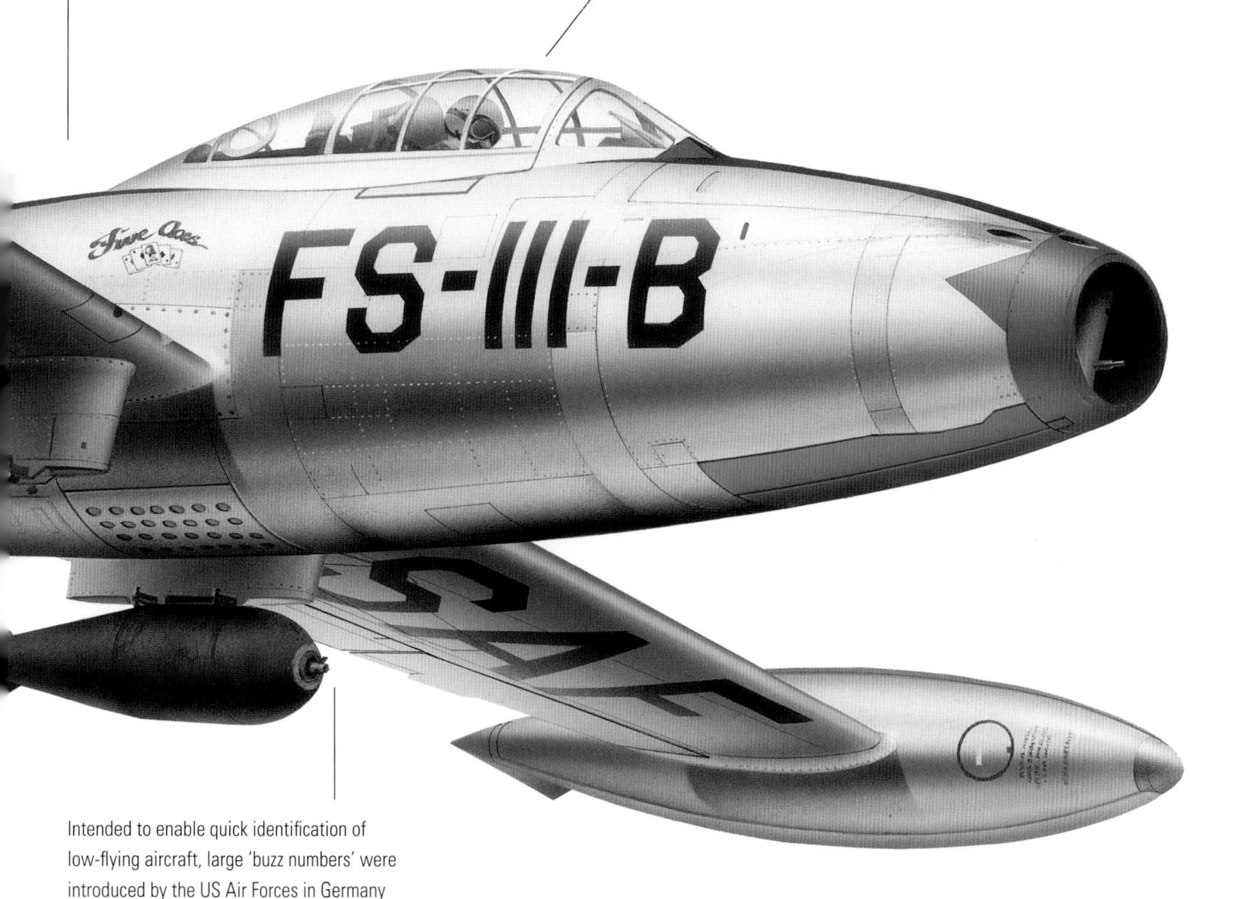

Intended to enable quick identification of low-flying aircraft, large 'buzz numbers' were introduced by the US Air Forces in Germany late in 1945 and were soon applied across the service. Comprising a two-letter or three-letter code (the first letter(s) indicating the aircraft's role and the last its type), plus the last three digits of its serial number, the 'buzz number' was generally applied in black on the aircraft's nose.

The Republic F-84 Thunderjet, which was to provide many of NATO's air forces with their initial jet experience, began life in the summer of 1944, when Republic Aviation's design team investigated the possibility of adapting the airframe of the P-47 Thunderbolt to take an axial-flow turbojet. This proved impractical, and in November 1944 the design of an entirely new airframe was begun around the General Electric J35 engine. The first of three XP-84 prototypes was completed in December 1945 and made its first flight on 28 February 1946. Three prototypes were followed by 15 YP-84As for the USAF. Delivered in the spring of 1947, they were later converted to F-84B standard. The F-84B was the first production model, featuring an ejection seat, six 12.7mm (0.50in) M3 machine guns and underwing rocket racks. Deliveries of the F-84B began in the summer of 1947 to the 14th Fighter Group, and 226 were built. The F-84C, of which 191 were built, was externally similar to the F-84B, but incorporated an improved electrical system and an improved bomb release mechanism. The next model

Below: **The Zero-Length Launcher (ZELL) concept, intended as a front-line dispersal measure in time of war, was first tested on the Republic F-84 Thunderjet, and later on the much heavier North American F-100 Super Sabre. The idea was soon abandoned.**

to appear, in November 1948, was the F-84D, which had a strengthened wing and a modified fuel system. Production totalled 151 aircraft. It was followed, in May 1949, by the F-84E, which in addition to its six 12.7mm (0.50in) machine guns could carry two 454kg (1000lb) bombs, or 32 rockets. The F-84G, which appeared in 1952, was the first Thunderjet variant to be equipped for flight refuelling from the outset. It was also the first USAF fighter to have a tactical nuclear capability.

The Thunderjet was widely used during the Korean War; although completely outclassed as a fighter by the MiG-15, it was very effective in the ground attack role. The F-84G made its appearance in Korea in 1952, and in the closing months of the war Thunderjets of the 49th and 58th Fighter-Bomber Wings carried out a series of heavy attacks on North Korea's irrigation dams, vital to that country's economy. The first target was the Toksan dam, a 700m (2300ft) earth and stone structure on the Potong river 20 miles north of Pyongyang. The dam was attacked in the afternoon of 13 May by 59 Thunderjets of the 58th FBW, armed with 454kg (1000lb) bombs, and the result seemed disappointing: apart from a slight crumbling of the structure, the dam still stood. The next morning, however, photographs brought back by an RF-80 revealed a scene of total destruction. During the night the pressure of water in the reservoir had caused the dam to collapse, sending a mighty flood down the Potong valley. Five square miles of rice crops had been swept away, together with 700 buildings; Sunan airfield was under water, and five miles of railway line, together with a two-mile stretch of the adjacent north–south highway, had been destroyed or damaged. In this one attack, the F-84s had inflicted more damage on the enemy's transport system than they had done in several weeks of interdiction work.

Above: **An F-84G Thunderjet in the colourful insignia of the 9th Fighter-Bomber Squadron, the 'Iron Knights'. The Thunderjet remained at the forefront of NATO's tactical nuclear strike force until it began to be replaced by the swept-wing F-84F Thunderstreak in 1955.**

Encouraged by this success, the American commander-in-chief, General Weyland, immediately authorized attacks on two more dams, at Chasan and Kuwonga. Next day 36 Thunderjets of the 58th FBW attacked the Chasan dam. The bomb-aiming, however, was poor; no direct hits were registered, and it was not until five 1000lb bombs were placed squarely on the target by a second Thunderjet strike on 16 May that the structure crumbled and the waters burst through to inundate a large area of rice and destroy a half-mile stretch of the nearby railway line. The Thunderjet's final missions in Korea were flown on 27 July, the very last day of hostilities, when the 49th and 58th FBWs attacked three airfields in the north.

Specification: Republic F-84G Thunderjet	
Type:	single-seat fighter-bomber
Powerplant:	one 2539kg (5600lb) thrust Wright J65-A-29 turbojet
Performance:	maximum speed 973km/h (605mph) at 1220m (4000ft); service ceiling 12,344m (40,500ft); range 1609km (1000 miles)
Weights:	empty 5200kg (11,460lb); maximum take-off 12,700kg (28,000lb)
Dimensions:	wing span 11.07m (36ft 4in); length 11.71m (38ft 5in); height 3.91m (12ft 10in); wing area 24.18m² (260 sq ft)
Armament:	six 12.7mm (0.50in) Browning M3 machine guns; provision for up to 1814kg (4000lb) of external stores

North American F-86 Sabre

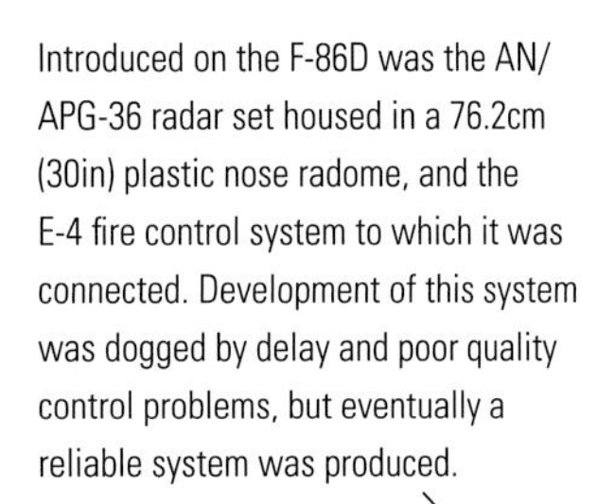

Introduced on the F-86D was the AN/APG-36 radar set housed in a 76.2cm (30in) plastic nose radome, and the E-4 fire control system to which it was connected. Development of this system was dogged by delay and poor quality control problems, but eventually a reliable system was produced.

The all-weather F-86D was originally intended to be a two-man aircraft. The back-seater would have been responsible for radar-controlled interception and navigation, but the performance limitations inherent in a two-seat design, plus the problems of the resulting reduction in fuel tank space, led to its abandonment.

A retractable rocket pack contained the F-86D's air-to-air armament of 24 70mm (2.75in) folding-fin aircraft rockets (FFAR). Known as 'Mighty Mouse' rockets, they were originally developed for use by the US Navy and were based on the German R4M rocket produced for the Messerschmitt 262.

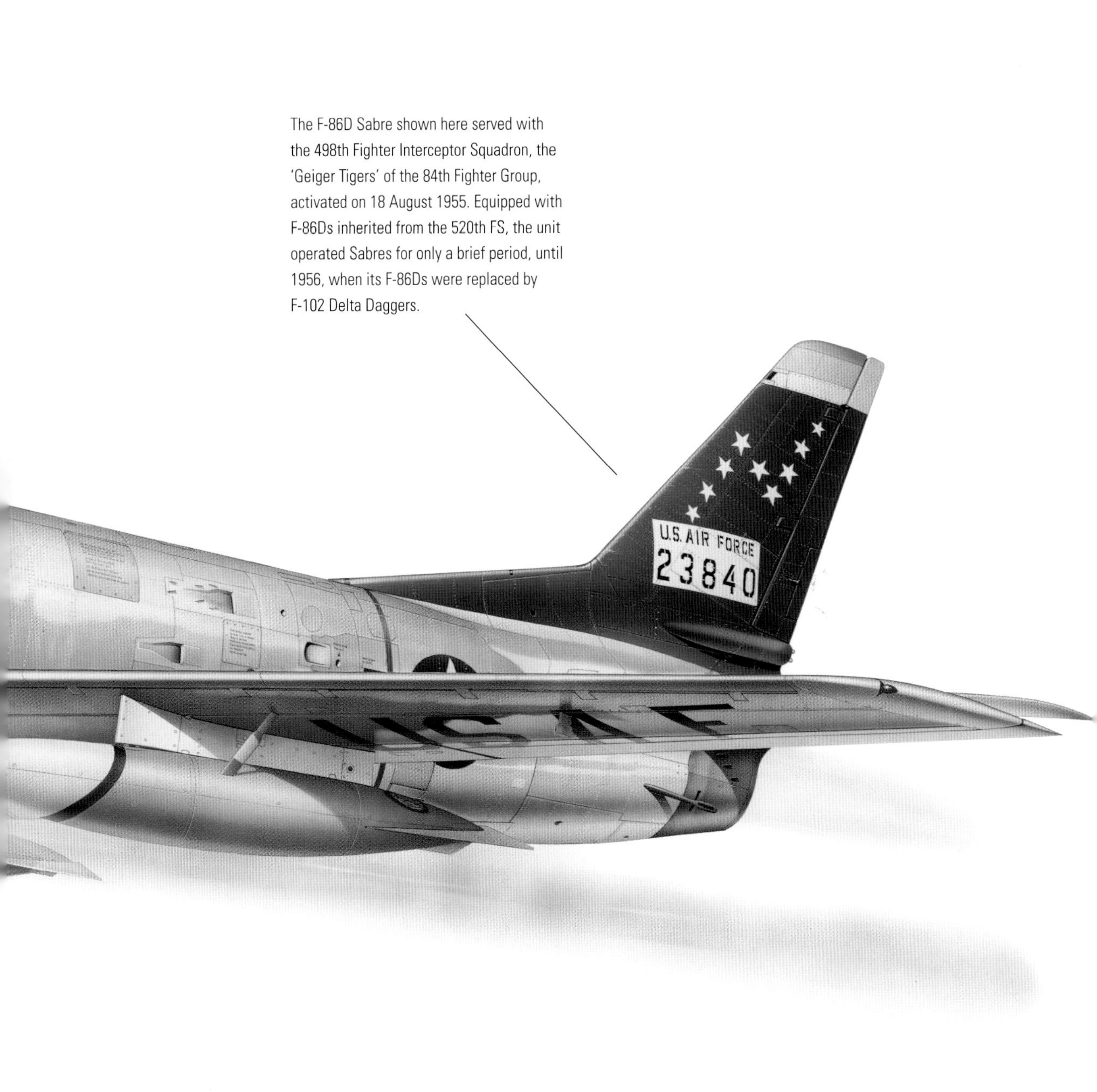

The F-86D Sabre shown here served with the 498th Fighter Interceptor Squadron, the 'Geiger Tigers' of the 84th Fighter Group, activated on 18 August 1955. Equipped with F-86Ds inherited from the 520th FS, the unit operated Sabres for only a brief period, until 1956, when its F-86Ds were replaced by F-102 Delta Daggers.

In 1944, before German advanced aeronautical research data became available, the USAAF issued specifications drawn up around four different fighter requirements, the first of which involved a medium-range day fighter that could also serve in the ground attack and bomber escort roles. This awakened the interest of North American Aviation, whose design team was then working on the NA-134, a projected carrier-borne jet fighter for the US Navy (which emerged as the FJ-1 Fury). The NA-134 was of conventional straight-wing design and was well advanced, so North American offered a land-based version to the USAAF under the company designation NA-140. On 18 May 1945 North American received a contract for the building of three NA-140 prototypes under the USAAF designation XP-86. A mock-up of the XP-86 was built and, in June 1945, was approved by the USAAF. There was, however, one worrying factor. According to North American's estimates, the XP-86 would have a maximum speed of 574mph at sea level, which fell short of the USAAF specification. Fortunately, it was at this point that material on German research into high-speed flight, in particular swept-wing designs, became available. North American obtained a complete Me 262 wing assembly and, after carrying out more than 1000 wind tunnel tests on it, decided that the swept wing was the answer to the XP-86's performance problems. The redesigned XP-86 airframe, featuring sweepback on all flying surfaces, was

Below: **The prototype XP-86 Sabre was flown for the first time by test pilot George S. Welch on 1 October 1947. Welch, a hero of Pearl Harbor, may well have exceeded Mach 1.0 on that day, two weeks before 'Chuck' Yeager became the first man to officially break the sound barrier.**

accepted by the USAAF on 1 November 1945 and received final approval on 28 February 1946. In December 1946 the USAAF placed a contract for an initial batch of 33 P-86A production aircraft, and on 8 August 1947 the first of two flying prototypes was completed, making its first flight under the power of a General Electric J35 turbojet. The second prototype, designated XF-86A, made its first flight on 18 May 1948, fitted with the more powerful General Electric J47-GE-1 engine, and deliveries of production F-86As began ten days later. The first operational F-86As were delivered to the 1st Fighter Group early in 1949. As yet, the F-86A was an aircraft without a name, and one of the 1st Fighter Group's acts was to sponsor a competition to find a suitable one. Seventy-eight names were submitted, and one stood out above the rest. On 4 March 1949 the North American F-86 was officially named the Sabre.

Production of the F-86A ended with the 554th aircraft in December 1950, a date that coincided with the arrival of the first F-86As in Korea with the 4th Fighter Wing. During the next two and a half years Sabres were to claim the destruction of 810 enemy aircraft, 792 of them MiG-15s. The next Sabre variants were the F-86C penetration fighter (which was redesignated YF-93A and which flew only as a prototype) and the F-86D all-weather fighter, which had a complex fire control system and a ventral rocket pack: 2201 were built, the F-86L being an updated version. The F-86E was basically an F-86A with power-operated controls and an all-flying tail; 396 were built before the variant was replaced by the F-86F, the major production version with 2247 examples being delivered. The F-86H was a specialized fighter-bomber armed with four 20mm (0.79in) cannon and capable of carrying a tactical nuclear weapon; the F-86K was essentially a simplified F-86D; and the designation F-86J was applied to the Canadair-built Sabre Mk.3. Most of the Sabres built by Canadair were destined for NATO air forces; the RAF, for example, received 427 Sabre Mk.4s. The Sabre Mk.6 was the last variant built by Canadair. The Sabre was also built under licence in Australia as the Sabre Mk.30/32, powered by a Rolls-Royce Avon turbojet. The total number of Sabres built by North American, Fiat and Mitsubishi was 6208, with a further 1815 produced by Canadair.

Specification: North American F-86E Sabre	
Type:	single-seat fighter
Powerplant:	one 2358kg (5200lb) thrust General Electric J47-GE-13 turbojet
Performance:	maximum speed 1086km/h (675mph) at 7620m (25,000ft); service ceiling 14,720m (48,300ft); range 1260km (765 miles)
Weights:	empty 5045kg (11,125lb); maximum take-off 7419kg (14,720lb)
Dimensions:	wing span 11.30m (37ft 1in); length 11.43m (37ft 6in); height 4.47m (14ft 8in); wing area 27.76m² (288 sq ft)
Armament:	six 12.7mm (0.50in) Colt-Browning machine guns; up to 907kg (2000lb) of underwing stores

Below: **This F-86A Sabre, 48-259, was flown in his first tour of operations by Captain (later Lt Col) James Jabara of the 4th FIW, who shot down 15 MiG-15s during the Korean War to become the second-ranking ace after Captain Joseph McConnell of the 51st FIW, who claimed 16.**

McDonnell F-101 Voodoo

The interceptor version of the Voodoo was fitted with a Hughes MG-13 fire control system, with a large and powerful radar in the nose. The system provided automatic intercept cues based on radar returns. In addition to the radar, the F-101B featured an infrared search and track (IRST) ball mounted above the nose.

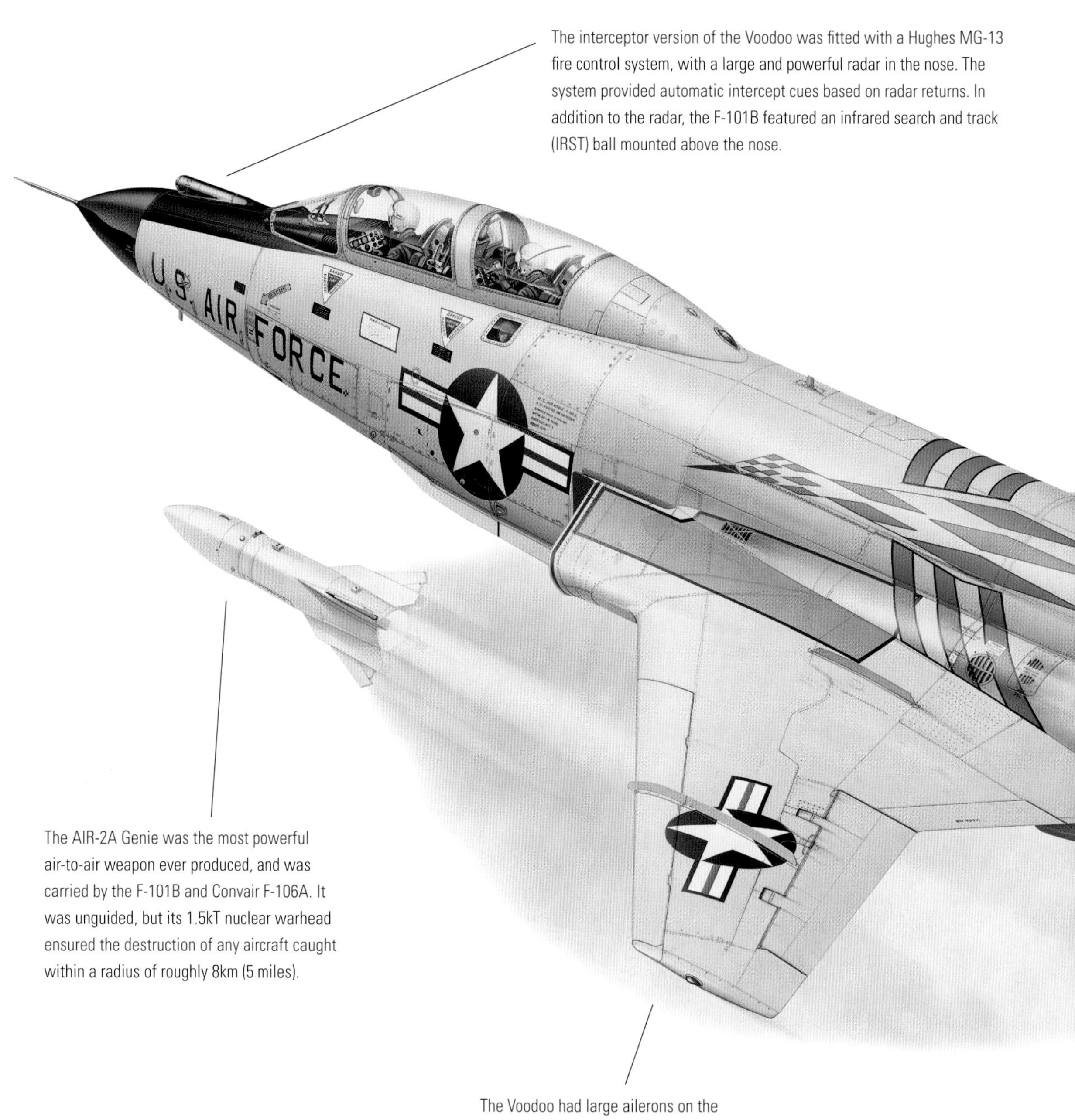

The AIR-2A Genie was the most powerful air-to-air weapon ever produced, and was carried by the F-101B and Convair F-106A. It was unguided, but its 1.5kT nuclear warhead ensured the destruction of any aircraft caught within a radius of roughly 8km (5 miles).

The Voodoo had large ailerons on the outboard section of the wing trailing edge, with split flaps inboard. The outer portion of the leading edge was hinged to act as a flap, so augmenting the lift from the small and thin wing.

The 60th Fighter Interceptor Squadron, Aerospace Defense Command, based its F-101Bs at L.G. Hanscom Field, Massachusetts. This aircraft, in common with the USAF's dedicated interceptor force at the time, is seen in 'ADC grey' scheme of gloss light grey with white undersides.

In 1946 the USAF Strategic Air Command issued a requirement for a so-called 'penetration fighter', intended primarily to escort the Convair B-36, or rather to sweep ahead of the bomber force and tear gaps in the enemy's fighter defences. One of the contenders was the McDonnell XF-88, prototype construction of which began in 1947 under a USAF contract. The prototype XF-88 was powered by two 1360kg (3000lb) thrust Westinghouse XJ34-WE-13 engines, mounted side-by-side and exhausting just aft of the wing trailing edge under a stepped up rear fuselage. This aircraft flew on 20 October 1948, and in 1950 it was followed by a second prototype fitted with XJ34WE-22 engines equipped with short afterburners that could boost the thrust to 1814kg (4000lb) thrust for combat manoeuvres. The XF-88 had a very thin wing swept at 35 degrees and spanning 11.79m (38ft 8in); length was just over 16.47m (54ft). The prototype XF-88 reached a maximum speed of 1030km/h (641 mph) at sea level and could climb to 10,670m (35,000ft) in four and a half minutes. Combat radius, however, was 1368km (850 miles), and operational ceiling was only 10,980m (36,000ft) The

Below: **The F-101B variant of the Voodoo equipped 16 squadrons of the USAF Air Defense Command, production running to 359 aircraft, and also equipped three Canadian air defence squadrons as the CF-101B, replacing the cancelled CF-105 Arrow.**

Above: **During the 1970s, McDonnell F-101B/F Voodoos equipped seven Air National Guard squadrons, including Minnesota's 179th Fighter Interceptor Squadron, which used the type from April 1971 until the winter of 1975.**

XF-88 development programme was cancelled in August 1950, when the USAF shelved its long-range heavy fighter plans, but the prototype – as the XF-88B – went on to have a useful life as a test bed for supersonic propellers.

In 1951 the USAF briefly resurrected its long-range escort fighter requirement as a result of the combat losses suffered in Korea by SAC's B-29s, and McDonnell used the XF-88 design as the basis for a completely new aircraft, lengthening the fuselage to accommodate two Pratt & Whitney J57-P-13 engines, giving it a top speed of over 1000mph and a ceiling of 52,000 feet, and increased fuel tankage. In its new guise it became the F-101A Voodoo, an aircraft that was to serve the USAF well for many years in the tactical support and reconnaissance role, even when the penetration fighter requirement was cancelled yet again.

The resurrected design was subjected to a number of changes, including the lengthening of the fuselage by over 4.00m (13ft) to accommodate extra fuel tankage, and the remodelled aircraft was designated YF-101A. The prototype flew on 29 December 1954, and although Strategic Air Command had long since abandoned the long-range escort fighter idea, the programme was taken over by Tactical Air Command, which saw the F-101 as a potential replacement for the Northrop F-89 Scorpion. The aircraft went into production as the F-101A, powered by two Pratt & Whitney J57-P-13 turbojets, and the 75 examples built equipped three squadrons of TAC. The next Voodoo variant, the two-seat F-101B, equipped 16 squadrons of Air Defense Command, and production ran to 359 aircraft. This version also equipped three Canadian air defence squadrons as the CF-101B, and formed a very important component of Canada's air defences; it replaced the very advanced Avro Canada CF-105 Arrow, which was cancelled on grounds of economy. The RCAF's CF-105Bs were normally held at five-minute readiness when on air defence alert. Most interceptions of Soviet reconnaissance aircraft were made by No 416 Squadron at Chatham, Ontario, by nature of its geographical location. The F-101C was a single-seat fighter-bomber version for TAC, entering service with the 523rd Tactical Fighter Squadron of the 27th Fighter Bomber Wing in May 1957. It equipped nine squadrons, but its operational career was relatively short-lived, as it was replaced by more modern combat types in the early 1960s. F-101C Voodoos were based at RAF Bentwaters, Suffolk, England, with the 81st Tactical Fighter Wing from 1958 to 1965, when the type was replaced by the F-4C Phantom. The Voodoo replaced the 81st TFW's F-84F Thunderstreaks, which were transferred to the Federal German Luftwaffe.

Specification: McDonnell F-101B Voodoo

Type:	two-seat long-range interceptor
Powerplant:	two 7664kg (16,900lb) thrust Pratt & Whitney J57-P-55 turbojets
Performance:	maximum speed 1965km/h (1221mph) at 12,190m (40,000ft); service ceiling 16,705m (54,800ft); range 2494km (1550 miles)
Weights:	empty 13,138kg (28,970lb); maximum take-off 23,763kg (52,400lb)
Dimensions:	wing span 12.09m (39ft 8in); length 20.54m (67ft 4in); height 5.49m (18ft 0in); wing area 34.10m² (368 sq ft)
Armament:	two AIR-2A Genie nuclear-tipped AAMs and four AIM-4C, -4D or -4G Falcon missiles, or six Falcon AAMs

Lockheed U-2

The camera compartment, or Q-Bay, was installed in the fuselage immediately aft of the cockpit and was accessed via two doors, one in the fuselage spine, the other in the belly. The main camera, known as the Model 73B (or simply the Type B), was a revolutionary piece of equipment, incorporating a system that damped out engine vibration and compensated for the motion of the aircraft so that blurring would be eliminated, or as near as possible.

The U-2's cockpit featured a manually operated canopy that hinged to one side in the same way as the F-104's and there was no ejection seat. The U-2 pilot worked in a unique environment. An unusual feature was the food heater, which prepared astronaut-type meals for consumption via a tube.

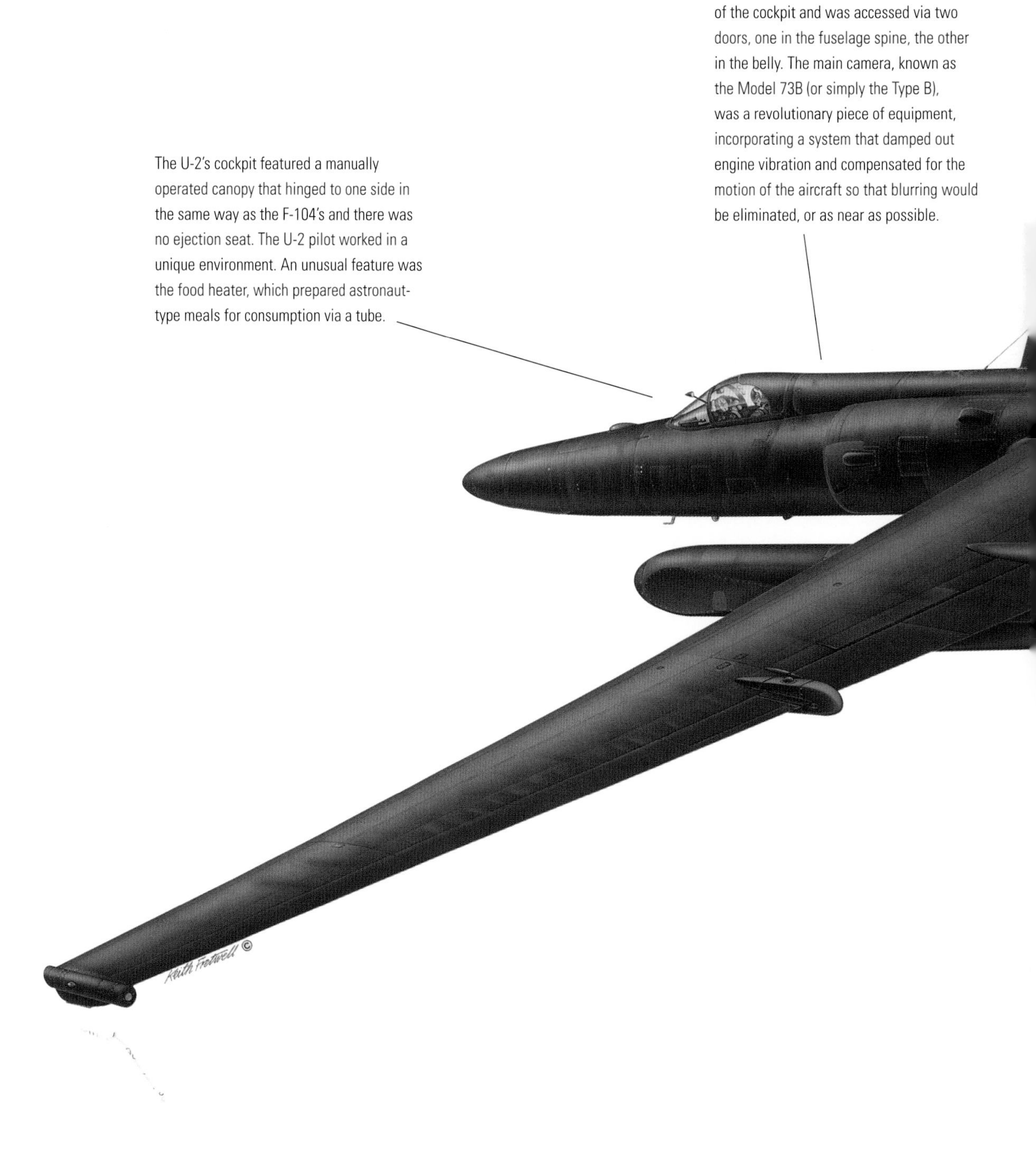

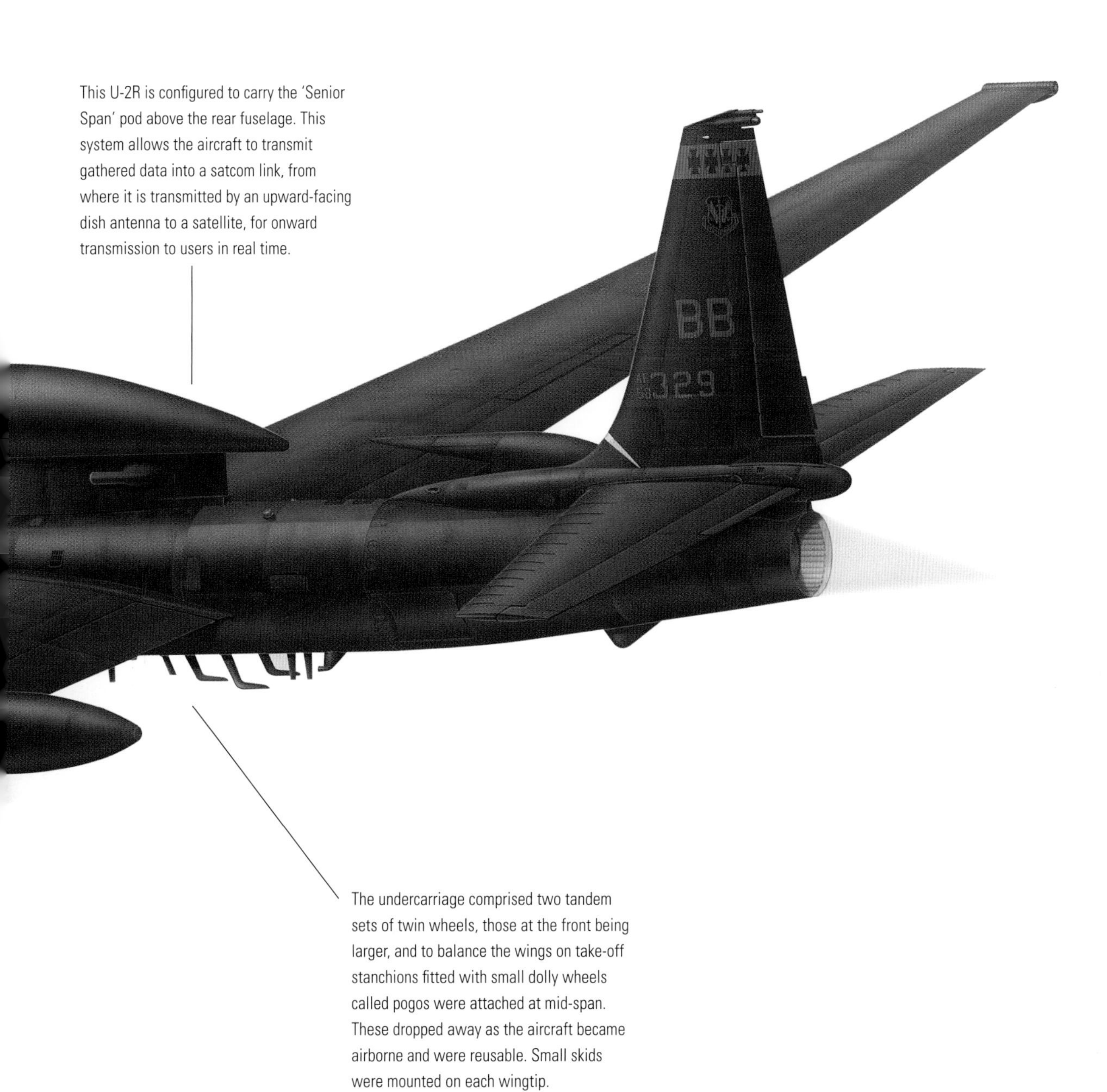

This U-2R is configured to carry the 'Senior Span' pod above the rear fuselage. This system allows the aircraft to transmit gathered data into a satcom link, from where it is transmitted by an upward-facing dish antenna to a satellite, for onward transmission to users in real time.

The undercarriage comprised two tandem sets of twin wheels, those at the front being larger, and to balance the wings on take-off stanchions fitted with small dolly wheels called pogos were attached at mid-span. These dropped away as the aircraft became airborne and were reusable. Small skids were mounted on each wingtip.

In March 1954 Clarence 'Kelly' Johnson, Lockheed's Chief Designer, presented a proposal for a high-altitude reconnaissance aircraft to the US Air Force, the Korean War having shown that existing types had a low survival factor in hostile airspace. Known as the Lockheed Model CL-282, the design was based on the fuselage and tail unit of a Lockheed F-104 Starfighter mated with a very high aspect ratio wing. The proposal was rejected on the grounds of the chosen engine; Johnson wanted the experimental General Electric J73, while the USAF favoured the proven Pratt & Whitney J57. The Air Force's caution was understandable; with long flights over hostile territory envisaged, engine reliability meant survival.

Undeterred, Johnson submitted the project to officials of the Central Intelligence Agency, and following a meeting with CIA Director Allan Dulles and the agency's Chief of Research and Development, Dr Joe Charyk, agreement was reached whereby Johnson would redesign the CL-282 around the Pratt & Whitney J57 turbojet, while still incorporating many of the F-104's features. Johnson indicated that Lockheed could build 20 aircraft plus spares for $22 million, and that a prototype would be ready within eight months of signing a contract.

On 9 December 1954 the CIA awarded Lockheed a development contract under the code name Project Aquatone, with funds for the airframes to be provided by the CIA and USAF funding for the engines. The prototype was produced under conditions of the utmost secrecy in Lockheed's Advanced Developments Projects Office, the engineering department of the Burbank factory known as the 'Skunk Works'. The name was derived from the 'Li'l Abner' cartoon strip character who brewed 'Kickapoo Joy Juice' in a shack from skunks, old boots and anything else that was handy, and was a hangover from 1943, when the XP-80 was being designed in makeshift workshops made from engine crates and circus tents adjacent to a foul-smelling plastics factory in Burbank.

Below: **The basic Lockheed U-2 design has been the subject of a good deal of development over the years. The last reconnaissance variant was the U-2R, from which the TR-1A battlefield surveillance aircraft was developed.**

Above: **This U-2R is fitted with the Senior Span pod above the fuselage. This configuration is used to transmit intelligence data from the aircraft's Senior Glass SIGINT suite using a satellite datalink, and was used extensively in operations over the former Yugoslavia.**

The aircraft that emerged, known simply as CIA Article 341, was virtually a jet-powered sailplane with a slim fuselage, long tapered wings and a tall fin and rudder. As the Lockheed U-2, it was destined to become the most controversial and politically explosive aircraft of all time.

The U-2 made its first flight in August 1955, an order for 52 production aircraft following quickly. Overflights of the USSR and Warsaw Pact territories began in 1956, and continued until 1 May 1960, when a Central Intelligence Agency pilot, Francis G. Powers, was shot down near Sverdlovsk by a Soviet SA-2 missile battery. U-2s were used to overfly Cuba during the missile crisis of 1962, one being shot down, and the type was also used by the Chinese Nationalists to overfly mainland China, all four aircraft being subsequently lost. U-2s also operated over North Vietnam in 1965/6. The last U-2 variant was the U-2R, but in 1978 the production line was reopened for the building of 29 TR-1A battlefield surveillance aircraft, developed from the U-2R. All TR-1As were redesignated U-2R in the 1990s.

Specification: Lockheed U-2R	
Type:	single-seat high altitude reconnaissance aircraft
Powerplant:	one 7711kg (17,000lb) thrust Pratt & Whitney J75-P-13B turbojet
Performance:	maximum speed 796km/h (495mph) at 12,200m (40,000ft); service ceiling 27,430m (90,000ft); range 4184km (2600 miles) with auxiliary tanks
Weights:	empty 7030kg (15,500lb); maximum take-off 18,730kg (41,300lb)
Dimensions:	wing span 31.39m (103ft 0in); length 19.13m (62ft 9in); height 4.88m (16ft 0in)
Armament:	none

Boeing B-52 Stratofortress

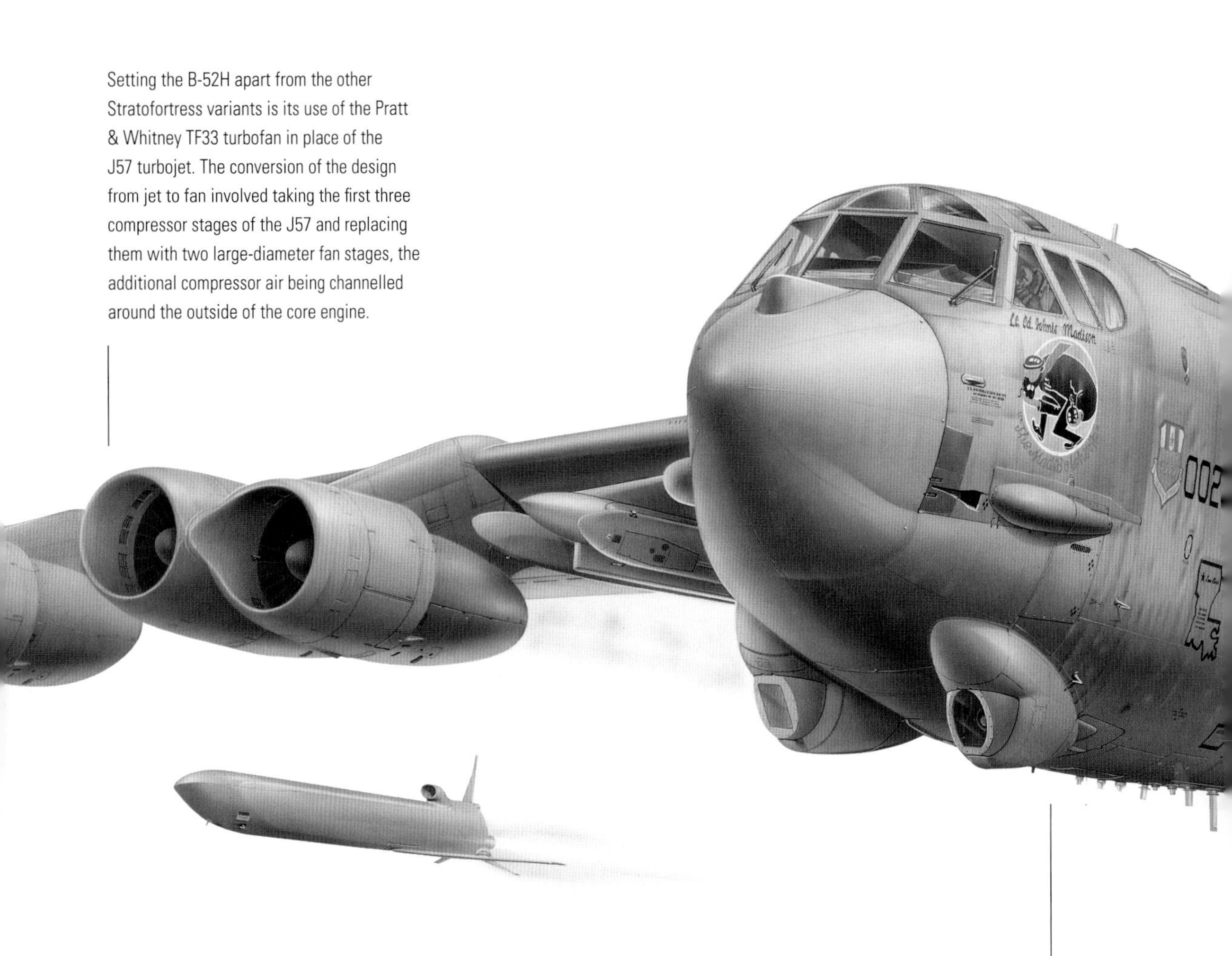

Setting the B-52H apart from the other Stratofortress variants is its use of the Pratt & Whitney TF33 turbofan in place of the J57 turbojet. The conversion of the design from jet to fan involved taking the first three compressor stages of the J57 and replacing them with two large-diameter fan stages, the additional compressor air being channelled around the outside of the core engine.

Seen here is a B-52H of the 20th Bomb Squadron, 2nd Bomb Wing, Eighth Air Force. Based at Barksdale AFB, Louisiana, the squadron is known as the 'Buccaneers'.

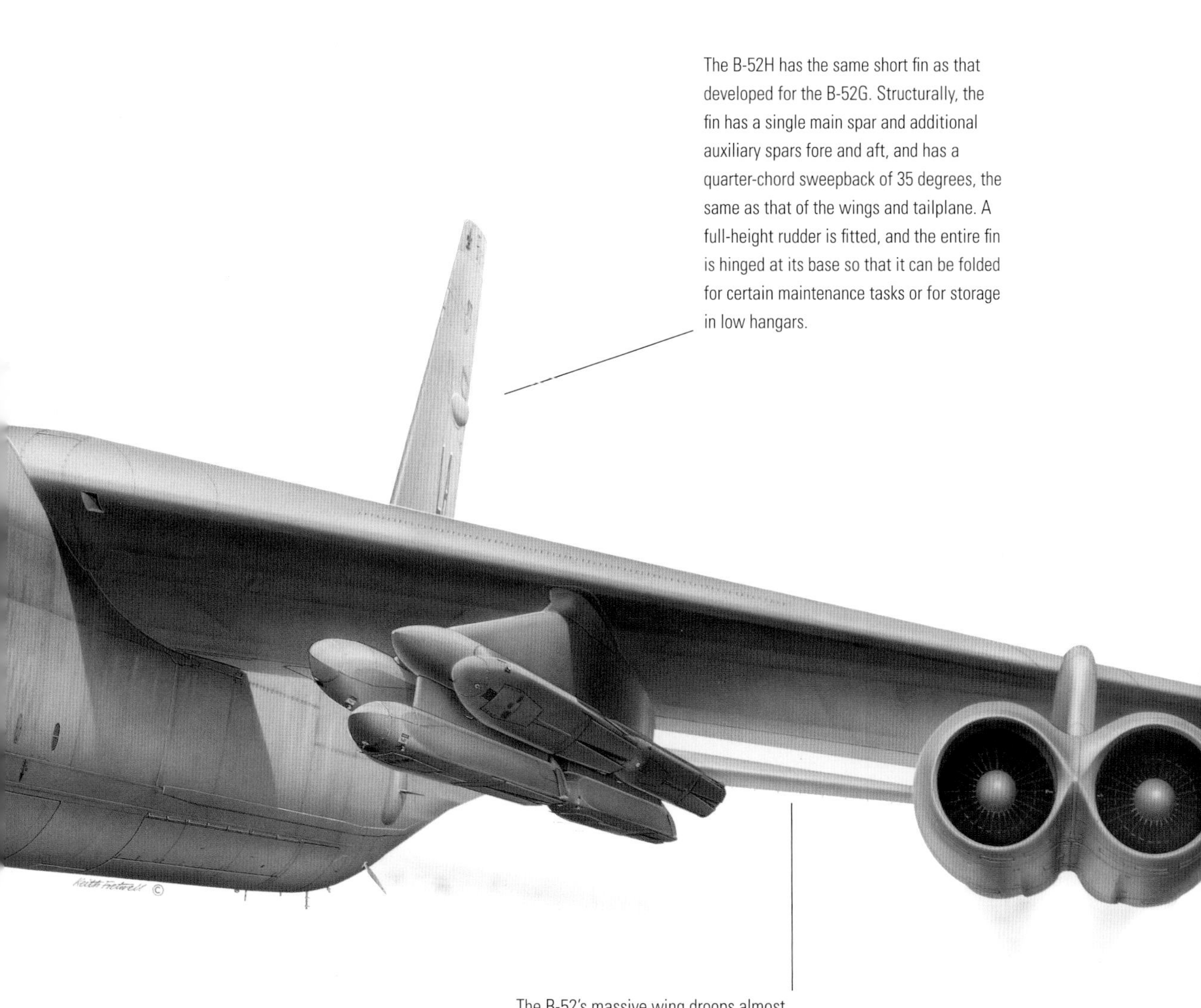

The B-52H has the same short fin as that developed for the B-52G. Structurally, the fin has a single main spar and additional auxiliary spars fore and aft, and has a quarter-chord sweepback of 35 degrees, the same as that of the wings and tailplane. A full-height rudder is fitted, and the entire fin is hinged at its base so that it can be folded for certain maintenance tasks or for storage in low hangars.

The B-52's massive wing droops almost low enough to touch the ground when the bomber is fully laden; each wingtip is equipped with an outrigger to prevent this from happening and also to provide stability during the take-off and landing rolls, the main undercarriage being situated under the fuselage.

During the dangerous years of the 1960s, the mighty Boeing B-52 was the symbol of America's awesome striking power. Few could have imagined that it would remain in first-line service half a century after the prototype first flew. It has been the core of the West's airborne strategic bomber forces ever since it entered service with the USAF Strategic Air Command in 1955, its operational career spanning almost all of the Cold War era. In addition, the B-52 has experienced the full range of technical and operational changes that have proven necessary to enable the strategic bomber to survive in an intensely hostile environment, particularly one dominated by sophisticated surface-to-air missiles.

The B-52 was the product of a USAAF requirement, issued in April 1946, for a new jet heavy bomber to replace the Convair B-36 in Strategic Air Command. Two prototypes were ordered in September 1949, the YB-52 flying for the first time on 15 April 1952 powered by eight Pratt & Whitney J57-P-3 turbojets. On 2 October 1952 the XB-52 also made its first flight, both aircraft having the same powerplant. The two B-52 prototypes were followed by three B-52As, the first of which flew on 5 August 1954. These aircraft featured a number of modifications and were used for extensive trials, which were still in progress when the first production B-52B was accepted by SAC's 93rd Bomb Wing at Castle AFB, California. Fifty examples were produced for SAC (including ten of the thirteen B-52As originally ordered, which were converted to B-52B standard) and it was followed on the production line by the B-52C, 35 of which were built. The focus of B-52 production then shifted to Wichita with the appearance of the B-52D, the first of which flew on 14 May 1956; 170 were eventually built. Following the B-52E (100 built) and the B-52F (89) came the major production variant, the B-52G.

The B-52G was the first aircraft to be armed with a long-range stand-off air-to-surface missile, the North American

Below: **The B-52G Stratofortress was developed to carry the GAM-77 Hound Dog air-to-surface missile, and was later adpated to carry a variety of other weapons, such as the Air-Launched Cruise Missile (ALCM). It was used operationally in the Gulf War, the Balkans and Afghanistan.**

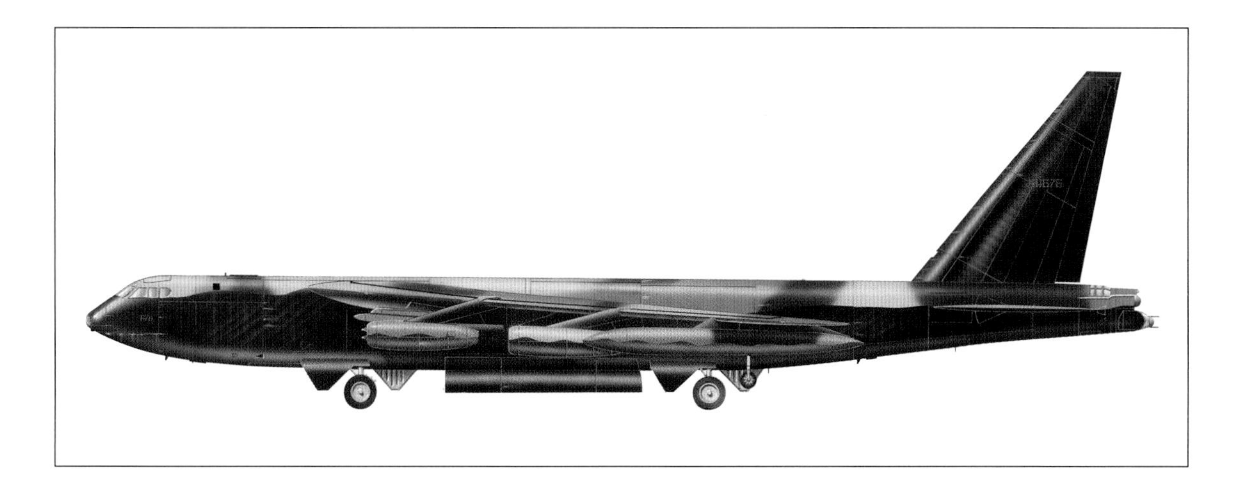

Above: **The B-52D Stratofortress, with its long range, high fuel capacity and ability to carry extra bombs in special underwing cells, was used in the bombing of Vietnam. Operating from Guam in the Pacific, the B-52 squadrons suffered heavy losses in attacks on targets in the Hanoi area.**

GAM-77 Hound Dog, a system designed to enhance the bomber's chances of survival. The missile was designed to carry a one-megaton warhead over a range of between 926 and 1297km (500 and 700nm) depending on the mission profile, and could operate between tree-top level and 16,775m (55,000ft) at speeds of up to Mach 2.1. All B-52Gs and, later, B-52Hs armed with the Hound Dog carried one pylon-mounted round under each wing. The Hound Dogs' turbojets were lit up during take-off, effectively making the B-52 a ten-engined aircraft, and were subsequently shut down, the missile's tanks being topped up from the parent aircraft. At the peak of the missile's deployment in 1962 there were 592 Hound Dogs on SAC's inventory, and it is a measure of the system's effectiveness that it remained in operational service until 1976. B-52G production totalled 193 examples, 173 of these being converted in the 1980s to carry 12 Boeing AGM-86B Air-Launched Cruise Missiles. The last version was the B-52H, which had been intended to carry the cancelled Skybolt air-launched IRBM but was modified to carry four Hound Dogs instead. The B-52 was also armed with the Boeing AGM-69 SRAM (Short-Range Attack Missile), the first being delivered to the 42nd Bomb Wing at Loring AFB, Maine, on 4 March 1972. The B-52 was capable of carrying twenty SRAMs, twelve in three-round underwing clusters and eight in the aft bomb bay, together with up to four Mk 28 thermonuclear weapons.

The B-52 was the mainstay of the West's airborne nuclear deterrent forces for three decades, but it was in a conventional role that it went to war, first over Vietnam, then in the Gulf War of 1991, and more latterly in support of NATO operations in the former Yugoslavia and anti-terrorist operations in Afghanistan. In all, 729 B-52 sorties were flown during the Linebacker II bombing offensive in Vietnam, and more than 15,000 tons of bombs dropped out of a total of 20,370 tons. Fifteen B-52s were lost to the SAM defences, and nine damaged. Thirty-four targets had been hit, and some 1500 civilians killed. Of the 92 crew members aboard the shot-down bombers, 26 were recovered by rescue teams, 29 were listed as missing, and 33 baled out over North Vietnam to be taken prisoner and later repatriated. As a result of the combat losses in Vietnam, the later variants of the B-52 were extensively rebuilt and upgraded with more advanced defensive avionics.

Specification: Boeing B-52D

Type:	six-place long-range strategic bomber
Powerplant:	eight 4535kg (10,000lb) thrust Pratt & Whitney J57-P-29WA turbojets
Performance:	maximum speed 1014km/h (630mph) at 7315m (24,000ft); service ceiling 16,765m (55,000ft); range with normal load 13,680km (8500 miles)
Weights:	empty 77,550kg (171,000lb); maximum take-off 204,120kg (450,000lb)
Dimensions:	wing span 56.39m (185ft); length 48.00m (157ft 5in); height 14.75m (48ft 4in); wing area 371.60m2 (4000 sq ft)
Armament:	remotely controlled tail barbette with four 12.7mm (0.50in) machine guns; up to 12,244kg (27,000lb) of conventional bombs; Mk.28 or Mk.43 nuclear free-falling weapons; two North American AGM-28B Hound Dog strategic stand-off missiles on underwing pylons

Lockheed F-104 Starfighter

General Electric's revolutionary M61 Vulcan cannon was first used in the F-104. A maximum of 725 rounds could be carried, with the cannon forming an important back-up system to the primary air-to-air armament of AIM-9 Sidewinder AAMs. All F-104 variants were Sidewinder-compatible, with two missiles normally being carried, either on under-fuselage rails or on wingtip launch rails if the wingtip fuel tanks were not fitted.

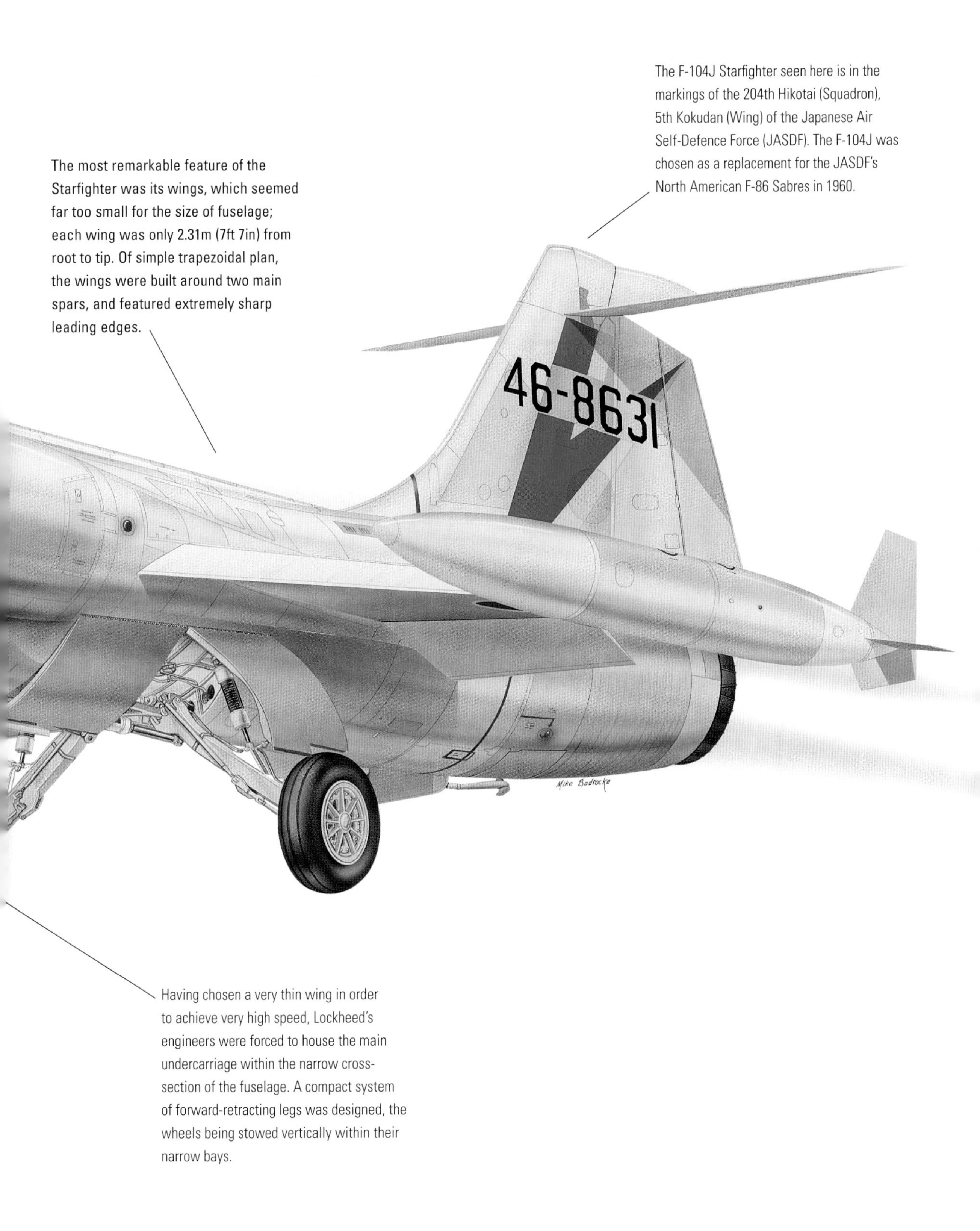
The F-104J Starfighter seen here is in the markings of the 204th Hikotai (Squadron), 5th Kokudan (Wing) of the Japanese Air Self-Defence Force (JASDF). The F-104J was chosen as a replacement for the JASDF's North American F-86 Sabres in 1960.
The most remarkable feature of the Starfighter was its wings, which seemed far too small for the size of fuselage; each wing was only 2.31m (7ft 7in) from root to tip. Of simple trapezoidal plan, the wings were built around two main spars, and featured extremely sharp leading edges.
46-8631
Mike Badrocke
Having chosen a very thin wing in order to achieve very high speed, Lockheed's engineers were forced to house the main undercarriage within the narrow cross-section of the fuselage. A compact system of forward-retracting legs was designed, the wheels being stowed vertically within their narrow bays.

Development of the F-104 was begun in 1951, when the lessons of the Korean air war were starting to bring about profound changes in combat aircraft design. A contract for two XF-104 prototypes was placed in 1953 and the first of these flew on 7 February 1954, only 11 months later. The two XF-104s were followed by 15 YF-104s for USAF evaluation, most of these, like the prototypes, being powered by the Wright J65-W-6 turbojet. The aircraft was ordered into production as the F-104A, deliveries to the USAF Air Defense Command beginning in January 1958. Because of its lack of all-weather capability the F-104A saw only limited service with Air Defense Command, equipping only two fighter squadrons. F-104As were also supplied to Nationalist China and Pakistan, and saw combat during the Indo-Pakistan conflict of 1969. The F-104B was a two-seat version, and the F-104C was a tactical fighter-bomber,

Left: **The F-104S Starfighter was a pure interceptor version, and was built under licence in Italy. The Italian Air Force was one of the last NATO air arms to operate the type, contributing to the aircraft's record as the longest-serving operational fighter in history.**

the first of 77 examples being delivered to the 479th Tactical Fighter Wing (the only unit to use it) in October 1958. Two more two-seat Starfighters, the F-104D and F-104F, were followed by the F-104G, which was numerically the most important variant. A single-seat multi-mission aircraft based on the F-104C, the F-104G had a strengthened structure and many equipment changes, including an upwards-ejecting Lockheed C-2 seat (earlier variants had a downward-ejecting seat). The first F-104G flew on 5 October 1960 and 1266 examples were produced up to February 1966, 977 by the European Starfighter Consortium and the remainder by Lockheed. Of these, the Luftwaffe received 750, the Italian Air Force 154, the Royal Netherlands Air Force 120 and the Belgian Air Force 99. The basically similar CF-104 was a strike-reconnaissance aircraft, of which 200 were built by Canadair for the RCAF. Canadair also built 110 more F-104Gs for delivery to the air forces of Norway, Nationalist China, Spain, Denmark, Greece and Turkey. Also similar to the F-104G was the F-104J for the Japan Air Self-Defence Force; the first one flew on 30 June 1961 and 207 were produced by Mitsubishi. The F-104S was an interceptor development of the F-104G, with provision for external stores, and was capable of Mach 2.4; 165 were licence-built in Italy.

Left: **Lockheed F-104G Starfighter of the Federal German Navy, which used the type predominantly in the anti-shipping strike role. The F-104G was replaced by the Panavia Tornado, after which many ex-German Starfighters were sold on to Greece or Turkey.**

Pilots were generally fond of the F-104. The cockpit was well designed and roomy, having all the instruments and switches conveniently located and affording an excellent view. Nosewheel steering made for easy ground manoeuvring, and was used up to 100 knots (185km/h, 115mph) during the take-off roll. In the air the Starfighter was remarkably stable, but the stick forces were heavy. Aerobatics in the rolling plane were quite conventional, but a great deal of sky was used up in the looping plane. A loop with maximum dry power was started from 500 knots (926km/h, 575mph) indicated airspeed and required about 3050m (10,000ft) of air space. Approach and landing presented no special problems. Constant speed approaches were flown, speed over the threshold being 175–205 knots (324–379km/h, 201–235mph), depending on fuel weight. The wheel brakes were very effective, brake protection being provided by an electrical generating device in each wheel, and the 5.5m (18ft) brake parachute could be used at speeds of up to 367km/h (228mph). An arrester hook was installed for emergency use.

Specification: Lockheed F-104G Starfighter	
Type:	single-seat multi-mission strike fighter
Powerplant:	one 7075kg (15,600lb) thrust General Electric J79-GE-11A turbojet
Performance:	maximum speed 1845km/h (1146mph) at 15,240m (50,000ft); service ceiling 15,240m (50,000ft); range 1740km (1081 miles)
Weights:	empty 6348kg (13,995lb); maximum take-off 13,170kg (29,035lb)
Dimensions:	wing span 6.63m (21ft 9in); length 16.66m (54ft 8in); height 4.09m (13ft 5in); wing area 18.22m² (196.10 sq ft)
Armament:	one 20mm (0.79in) General Electric M61A-1 cannon; Sidewinder AAMs on wing or fuselage stations; up to 1814kg (4000lb) of ordnance, including Bullpup ASMs

Grumman A-6 Intruder

The A-6's pilot and navigator sat on Martin-Baker GRU-5B or GRU-7 ejection seats. The seats, which could recline for comfort on long missions, were not positioned side-by-side, the navigator/WSO (Weapon Systems Officer) being seated below and aft of the pilot.

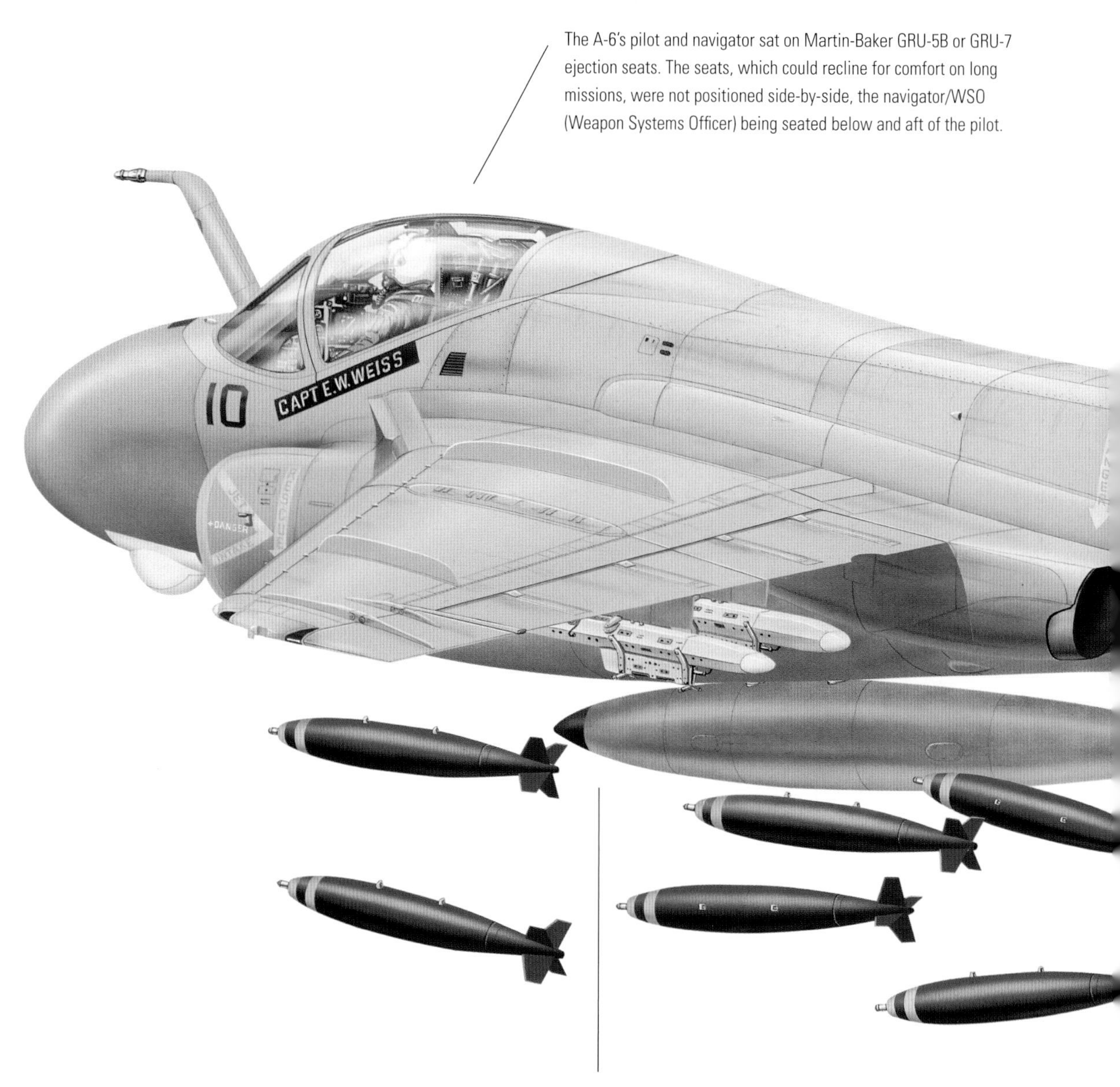

The A-6 carried its stores on the centreline and four underwing pylons. MERS (Multiple Ejector Racks) were universally used, these allowing the carriage of six small bombs on each, although in practice this was only applicable to the outboard pylons, the inboard racks being restricted to five for undercarriage retraction clearance. Large weapons such as SLAM/Harpoon missiles and nuclear weapons were usually carried on the inboard pylons.

The bulged fairing near the top of the fin housed antennae for the self-protection system. ALQ-126 DECM and ALR-67 threat warning receiver aerials were also mounted there. Other aerials were located in the wing root leading edge.

The A-6E pictured here belonged to VMA(AW)-121, which came under the control of MAG-11 at El Toro, California, until 11 May 1990, when it was redesignated VMFA(AW)-121. This was the first of five Marine A-6 squadrons to give up their aircraft over Fiscal Years 1990–95 as the Corps, along with the Navy, relinquished the Intruder type.

Designed specifically as a carrier-based low-level attack bomber with the ability to deliver both nuclear and conventional warloads with pinpoint accuracy in all weathers, the Grumman A-6 was one of 11 competitors in a US Navy design contest of 1957, and was selected as the winner in December that year. The A-6A prototype flew on 19 April 1960 and the first operational aircraft entered service with Attack Squadron VA-42 on 1 February 1963. The last delivery took place in December 1969, by which time 488 had been built. The A-6A saw extensive action over Vietnam, working round the clock and performing combat missions that were far beyond the capability of any other aircraft until the advent of the F-111, and also participated in later actions, such as the strike on Libya in April 1986. The next variant was the EA-6A electronic warfare aircraft, 27 examples of which were produced for the US Marine Corps; this was followed by the EA-6B Prowler, with advanced avionics and a longer nose section to accommodate two extra ECM specialists. The EA-6B Prowler is included in every US aircraft carrier deployment. Its primary mission is to protect fleet surface units and friendly strike aircraft by jamming hostile radars

Above: **The aircraft seen here is the KA-6D tanker version of the Intruder. These conversions of the A-6A had most of their bombing and weapon systems removed, including the radar. The wings and rear fuselage were strengthened.**

and communications. As a result of restructuring the US Department of Defense's assets in the mid-1990s, it was decided to replace the General Dynamics EF-111A with the EA-6B. Five new EA-6B squadrons were activated, four of them charged with supporting USAF Expeditionary Aerospace Wings operating on UN or NATO tasks overseas. As was the case with the EF-111A, the heart of the Prowler is the AN/ALQ-99 Tactical Jamming System. The Prowler can carry five jamming pods, one belly-mounted and the others on wing stations; each pod is integrally powered and houses two jamming transmitters that cover seven frequency bands.

The EA-6B has been the subject of numerous upgrades and is expected to remain in service until about 2010, after a career of more than 40 years in support of USN, USMC and USAF strike forces worldwide. The aircraft played a key part in suppressing Iraqi radar defence systems during the Gulf War. US Defense Department planning includes a scheme whereby the Prowler would be replaced by the F/A-18G 'Growler', an F/A-18E/F Hornet modified for escort and close-in jamming. Stand-off jamming would be provided by Air Force EB-52s and EB-1s, or by unmanned vehicles.

The last basic attack variant of the Intruder was the A-6E, which first flew in February 1970; total procurement orders called for 318 A-6Es, including 119 converted from A-6As. Other conversions of the basic A-6A were the A-6C, with enhanced night attack capability, and the KA-6D flight refuelling tanker. The close of 1996 saw the end of the A-6's 31-year combat career with the US Navy, the type being deemed too expensive to operate in a service forced to cut back on its front-line types and one whose deep-strike mission had become less important in the post-Cold War world. In the winter of 1996/7 the last Atlantic and Pacific Fleet A-6 squadrons made their final cruises before retirement. Many of the surviving East Coast examples were dumped into the Atlantic Ocean, off the Florida coast, to form an artificial reef. Today, only the EA-6B Prowler remains in service, operated by both the USN and USMC.

Left: **Grumman A-6 Intruder of Navy Attack Squadron VA-165. The Intruder gave splendid service in Vietnam, working round the clock and performing combat missions that were far beyond the capability of any other combat aircraft until the advent of the F-111.**

Specification: Grumman A-6A Intruder	
Type:	two-seat all-weather strike aircraft
Powerplant:	two 4218kg (9300lb) Pratt & Whitney J52-P-8A turbojets
Performance:	maximum speed 1043km/h (648mph) at sea level; service ceiling 14,480m (47,500ft); range 1627km (1011 miles) with full weapon load
Weights:	empty 12,130kg (26,746lb); maximum take-off 27,397kg (60,400lb)
Dimensions:	wing span 16.15m (53ft 0in); length 16.64m (54ft 7in); height 4.93m (16ft 2in); wing area 49.13m^2 (528.9 sq ft)
Armament:	five external hardpoints for up to 8165kg (18,000lb) of ordnance

McDonnell Douglas F-4 Phantom II

The prototype F4H-1s were fitted with APQ-50 radar, as used in the Douglas F4D Skyray, but the developed APQ-72 was fitted to the first production aircraft. From the nineteenth F4H-1, an 81cm (32in) antenna was fitted, dramatically altering the look of the Phantom while considerably increasing the radar's range.

The prominent fairing under the radome of the F-4B housed the AAA-4 infrared sensor. This required range data from the radar, but could then be used to track targets passively.

The Phantom proved itself many times over in combat with MiG-17s and MiG-21s over North Vietnam. This particular aircaft, an F-4B of VF-111 'Sundowners' (USS *Coral Sea*) was flown by Lt Garry L. Weigand (pilot) and Lt (jg) William C. Freckleton (RIO) when they shot down a MiG-17. Note the kill marking on the intake splitter plate.

Whether flown from land or ship, the Phantom employed a no-flare landing, which was more akin to a controlled crash. To be able to withstand this considerable battering required an immensely strong undercarriage. That of the F-4 was designed to routinely handle sink rates of up to 6.7m (22ft) per second.

Right: **The Phantom was developed for the defence suppression role as the F-4G Wild Weasel, carrying the Shrike or HARM anti-radar missiles. The requirement arose during the Vietnam War, when the USAF suffered significant losses to SA-2 surface-to-air missiles.**

One of the most potent and versatile combat aircraft ever built, the McDonnell (later McDonnell Douglas) F-4 Phantom II stemmed from a 1954 project for an advanced naval fighter designated F3H-G/H. The designation was changed to F4H-1, and later to F-4A. The F-4B was a slightly improved version with J79-GE-8 engines. The first fully operational Phantom squadron, VF-114, commissioned with F-4Bs in October 1961, and in June 1962 the first USMC deliveries were made to VMF(AW)-314. Total F-4B production was 649 aircraft. Deliveries of the F-4C to the USAF began in 1963, 583 aircraft being built. The RF-4B and RF-4C were unarmed reconnaissance variants for the USMC and USAF, while the F-4D was basically an F-4C with improved systems and redesigned radome. The major production version was the F-4E, 913 of which were delivered to the USAF between October 1967 and December 1976. F-4E export orders totalled 558. The RF-4E was the tactical reconnaissance version. The F-4F (175

Specification: McDonnell F-4E Phantom II	
Type:	two-seat fighter/attack aircraft
Powerplant:	two 8117kg (17,900lb) thrust General Electric J79-GE-17 turbojets
Performance:	maximum speed 2390km/h (1485mph) at altitude; service ceiling 18975m (62,250ft); range 2817km (1750 miles) on internal fuel
Weights:	empty 12,700kg (28,000lb); maximum take-off 26,303kg (58,000lb)
Dimensions:	wing span 11.71m (38ft 5in); length 17.75m (58ft 3in); height 4.95m (16ft 3in); wing area 49.24m² (530 sq ft)
Armament:	one 20mm (0.79in) M61A-1 Vulcan cannon and four AIM-7 Sparrow AAMs recessed under fuselage; up to 5886kg (12,980lb) of ordnance and stores on underwing pylons

built) was a version for the Luftwaffe, intended primarily for the air superiority role but retaining multi-role capability, while the F-4G Wild Weasel was the F4E modified for the suppression of enemy defence systems. The successor to the F-4B in USN/USMC service was the F-4J, which possessed greater ground attack capability; the first of 522 production aircraft was delivered in June 1976.

The first foreign nation to order the Phantom was Great Britain, the British aircraft being powered by Rolls-Royce RB.168-25R Spey 201 engines. Versions for the Royal Navy and the RAF were designated F-4K and F-4M respectively. Fifty-two F-4Ks were delivered to the RN in 1968/9 and these were progressively handed over to the RAF with the run-down of the RAF's fixed wing units, becoming the Phantom FG.1 in RAF service; the FG.1 was used in the air defence role. The RAF's own version, the F-4M Phantom FGR.2, equipped 13 air defence, strike and reconnaissance squadrons. By 1978 all the RAF's F-4M Phantoms, of which 118 were delivered, were assigned to air defence, replacing the Lightning in this role. Other foreign customers for the Phantom included the Imperial Iranian Air Force, which received some 200 F-4Es and 29 RF-4Es; the surviving aircraft, under new management, saw combat during the long-running war between Iran and Iraq in the 1980s. Israel also received over 200 F-4Es between 1969 and 1976, these aircraft seeing considerable action during the Yom Kippur war of 1973. F-4D Phantoms were delivered to the Republic of Korea Air Force as a temporary measure, pending the arrival of Northrop F-5As.

Left: **The Royal Navy was the first overseas customer for the Phantom, No 892 Squadron of the Fleet Air Arm operating the FG.Mk.1 version from the aircraft carrier HMS *Ark Royal*. When the latter was decommissioned in 1978 the aircraft were passed to the RAF.**

The Japanese Air Self-Defence Force equipped five squadrons with 140 Phantom F-4EJs, most of which were built under licence, and the RAAF leased 24 F-4Es in 1970. Phantoms were delivered to Spain, Greece and Turkey, so that by the mid-1970s several key NATO air forces were standardized on the type.

Lockheed SR-71A Blackbird

Early problems in the SR-71A programme prompted the development of a new high-flashpoint fuel. Originally known as PF-1 and later designated JP-7, this fuel would only ignite at high temperatures, reducing the risk of inadvertent fires caused by the high operating temperatures encountered at high Mach numbers.

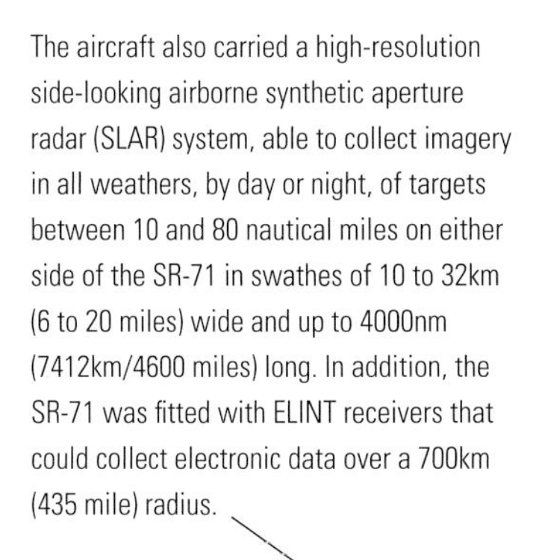

The aircraft also carried a high-resolution side-looking airborne synthetic aperture radar (SLAR) system, able to collect imagery in all weathers, by day or night, of targets between 10 and 80 nautical miles on either side of the SR-71 in swathes of 10 to 32km (6 to 20 miles) wide and up to 4000nm (7412km/4600 miles) long. In addition, the SR-71 was fitted with ELINT receivers that could collect electronic data over a 700km (435 mile) radius.

A complex method of environmental control comprised two air cycle systems that provided heating and cooling air for the cockpits and other aircraft systems. Three oxygen converters (one of which was a back-up) were provided for the crew's needs, and emergency oxygen supplies were incor-porated into the ejection seat survival kit.

The SR-71's main optical sensors included two 48-inch focal length cameras capable of photo-mapping terrain on either side of the aircraft's flight path over a distance of between 1544km (833nm) and 3000km (1619nm). In the nose, an optical bar camera (OBC) was designed to take long-range panoramic oblique photographs over hostile frontiers. With this camera, an SR-71 could photograph a strip of land between 2735km (1478nm) and 5421km (2930nm) long. At its normal operational altitude, and using more than one of its photographic systems, an SR-71 was capable of photographing 155,340km^2 (60,000 square miles) of territory in one hour.

On 25 July 1964 US President Lyndon B. Johnson lifted a small corner of the veil that shrouded one of the most secret programmes in the history of military aviation. Johnson announced that: 'The SR-71 aircraft will fly at more than three times the speed of sound. It will operate at altitudes in excess of 80,000 feet. It will use the most advanced observation equipment of all kinds in the world. The aircraft will provide the strategic forces of the United States with an outstanding long-range reconnaissance capability. The system will be used during periods of military hostilities and in other situations in which the United States military forces may be confronting foreign military forces ...'

The President had got everything right except one point. The aircraft was actually designated RS (Reconnaissance System) 71, but some official decided that it was easier to

Below: **In 1986 NASA's SR-71B was used to requalify USAF pilots for the resurrected Blackbird programme, the aircraft being brought out of retirement to fly further missions on behalf of NATO. Here, the two-seat trainer is eased in behind a KC-135 tanker during a training flight.**

rename the aircraft than inform Johnson that he had made a mistake.

Work on the SR-71 system began in 1959, when a team led by Clarence L. Johnson, Lockheed's Vice-President for Advanced Development Projects, embarked on the design of a radical new aircraft to supersede the Lockheed U-2 in the strategic reconnaissance role. Designated A-12, the new machine took shape in conditions of the utmost secrecy in the highly restricted section of the Lockheed Burbank plant, the so-called 'Skunk Works', and seven aircraft had been produced by the summer of 1964, when the project's existence was revealed. By that time, the A-12 had already been tested extensively at Edwards AFB, reaching speeds of more than 2000mph at heights of over 70,000 feet. Early flight tests were aimed at assessing the A-12's suitability as a long-range interceptor, and the experimental interceptor version was shown to the public at Edwards AFB in September 1964, bearing the designation YF-12A.

Two YF-12As were built, but the interceptor project was abandoned. Work on the strategic reconnaissance variant, however, went ahead and the prototype SR-71A flew for the first time on 22 December 1964. The first aircraft to be assigned to Strategic Air Command, an SR-71B two-seat trainer (61-7957) was delivered to the 4200th SRW at Beale AFB, California, on 7 January 1966. The 4200th SRW had been activated a year earlier, and by the time the first SR-71 was delivered selected crews had already undergone a comprehensive training programme on Northrop T-38s, eight of which were delivered to Beale from July 1965.

On 25 June 1966, with SR-71 deliveries continuing, the 4200th SRW was redesignated the 9th SRW, its component squadrons becoming the 1st and 99th SRS. In the spring of 1968, because of the growing vulnerability of the U-2 in a SAM environment, it was decided to deploy four SR-71s to Kadena Air Base, Okinawa, for operations over Southeast Asia. This deployment, known as Giant Reach, was on a 70-day TDY basis, with crews rotating between Beale and Kadena. The aircraft remained in situ and formed the nucleus of Detachment One of the 9th SRW. The first SR-71 mission over Vietnam was flown in April 1968, with up to three missions per week being flown thereafter.

SR-71 operations from the United Kingdom began on 20 April 1976, when 64-17972 arrived at RAF Mildenhall on a ten-day TDY. In the years that followed SR-71 deployments at RAF Mildenhall became a regular feature, the UK-based aircraft operating as Detachment Four, 9th SRW. (Detachment Four was originally a U-2 detachment, but the U-2s moved to RAF Alconbury in the early 1980s.) Two SR-71s were stationed in the UK at any one time, the aircraft flying stand-off surveillance missions over the Soviet Arctic, the Baltic and the Mediterranean. On 15 and 16 April 1986, two SR-71As of Detachment Four (64-17960 and 64-17980) carried out post-strike reconnaissance following the attacks on Libya (Operation Eldorado Canyon) by UK-based F-111s and US Navy aircraft; both SR-71s were used on each occasion.

Below: **The Lockheed SR-71 was operated by the 9th Strategic Reconnaissance Wing, which positioned detachments at various locations around the world. Two aircraft were normally deployed to these locations at any one time, one acting as a backup.**

Specification: Lockheed SR-71A	
Type:	two-seat strategic reconnaissance aircraft
Powerplant:	two 14,740kg (32,500lb) Pratt & Whitney JT11D-20B turbojets
Performance:	maximum speed 3220km/h (2000mph) at 24,385m (80,000ft); service ceiling 24,385m (80,000ft); range 4800km (2983 miles)
Weights:	empty 30,612kg (67,500lb); maximum take-off 78,000kg (172,000lb)
Dimensions:	wing span 16.94m (55ft 7in); length: 32.74m (107ft 5in); height 5.64m (18ft 6in); wing area 149.10m² (1605 sq ft)
Armament:	none

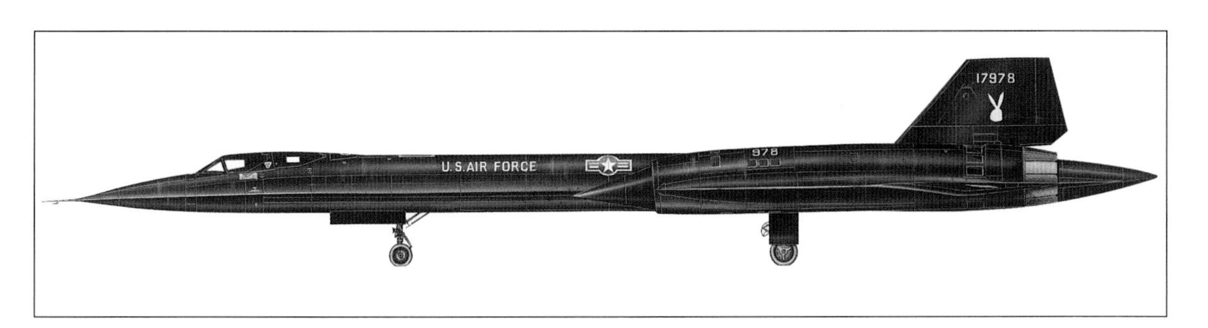

Northrop F-5 Tiger

This F-5E Tiger II is in the markings of the Brazilian Air Force's 1 Escadron, 1 Grupo de Aviação Caza; the 'fighting ostrich' insignia was originally used when the unit flew P-47D Thunderbolts with the USAAF's 350th Fighter Group in Italy during World War II.

Despite the small size of the aircraft the F-5's cockpit is very roomy, with ample storage space for maps. Pilots are particularly enthusiastic about the layout of switches and instruments, which is excellent. The F-5 is a very pleasant aircraft to fly, with no serious vices.

The original Brazilian F-5Es are almost invariably fitted with the detachable flight refuelling probe beneath the starboard canopy rail, allowing the aircraft to refuel at a rate of up to 907kg (2000lb) per minute. A small spotlight is fitted in the shoulder fairing at the base of the sloping downpipe, allowing the pilot to illuminate the probe tip and refuelling basket during night refuelling.

The F-5E is armed with a pair of 20mm (0.79in) Pontiac (Colt-Browning) M39A-2 cannon, with up to 280 rounds per gun stored in ammunition boxes below the barrels. The cannon each have a rate of fire of 1500 rounds per minute.

The Northrop N156 was conceived as a relatively cheap and simple aircraft capable of undertaking a variety of tasks. At the end of 1958 Northrop received a Department of Defense contract for three prototypes, the first of which flew on 30 July 1959, powered by two General Electric YJ85-GE-1 turbojets, and exceeded Mach One on its maiden flight. After nearly three years of intensive testing and evaluation, it was announced on 25 April 1962 that the N156 had been selected as the new all-purpose fighter for supply to friendly nations under the Mutual Aid Pact, and the aircraft entered production as the F-5A Freedom Fighter, the first example flying in October 1963. The F-5A entered service with USAF Tactical Air Command in April 1964. The first overseas customer was the Imperial Iranian Air Force, which formed

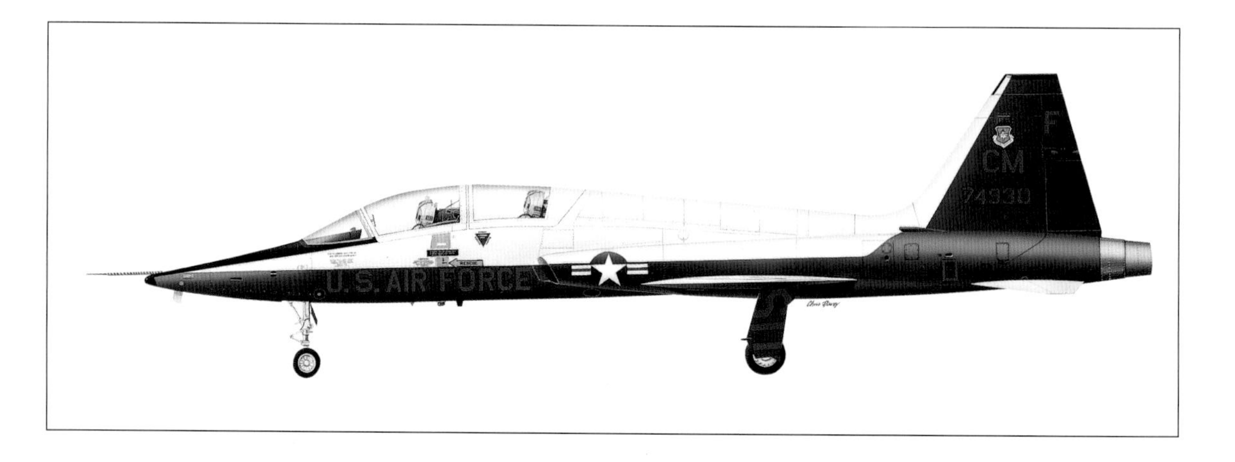

Left: **The Northrop F-5 has seen widespread service with NATO air forces. In the USAF, the F-5E Tiger II was used in the role of aggressor, bearing a strong resemblance to Russia's MiG-21 in terms of general configuration and all-round performance.**

the first of seven F-5A squadrons in February 1965. The Royal Hellenic Air Force also received two squadrons in 1965, and Norway received 108 aircraft from 1967, these being fitted with arrester hooks and rocket assisted take-off for short field operations. Between 1965 and 1970 Canadair built 115 aircraft for the Canadian Armed Forces as CF-5A/Ds, these using Orenda-built J85-CAN-15 engines. Other nations using the type were Ethiopia, Morocco, South Korea, the Republic of Vietnam, Nationalist China, the Philippines, Libya, the Netherlands, Spain, Thailand and Turkey. An improved version, the F-5E Tiger II, was selected in November 1970 as a successor to the F-5A series. It served with a dozen overseas air forces, and also in the 'aggressor' air combat training role with the USAF. The RF-5E TigerEye is a photo-reconnaissance version.

In the 1980s Northrop pinned a great deal of hope on the potential of an advanced version of the F-5, the F-20 Tigershark, and in the YF-23 advanced tactical fighter, which was in competition with the Lockheed F-22. Although a superb aircraft, the F-20 attracted no customers, and the F-22 was selected in preference to the YF-23, which left Northrop short of military contracts. The company therefore began design work on an upgraded version of the F-5 for the potential market of current operators. A number of other companies were already offering upgrades, but Northrop, which was not only the original equipment manufacturer but which also had experience of advanced avionics on the B-2 and F/A-18, saw itself as the logical choice.

Initial improvements were conducted under a structural upgrade programme funded by the USA, but in 1993 Northrop embarked on a much more ambitious programme using an F-5E 'borrowed back' from the US Navy. With a new avionics suite, including APG-66(V) radar, Honeywell ring laser gyro, AlliedSignal mission computer and display processor, an F-16 type GEC HUD and HOTAS controls plus a Martin-Baker Mk 10LF, the F-5E Tiger IV first flew on 20 April 1995. To accommodate the radar, one of the two cannon was removed and the radome lengthened. Bristol Aerospace, CASA and Samsung were selected as strategic partners and six months of trials followed. Now offered with various other avionics and electronic warfare suites, Tiger upgrades have been initiated by several air forces including Brazil, Chile, Indonesia and Singapore.

Left: **The Northrop T-38A Talon two-seat trainer formed the basis for a family of aircraft which also included the F-5A Freedom Fighter, to which it bears a strong resemblance. The T-38 proved highly successful in service with the US and foreign air forces.**

Specification: Northrop F-5A Tiger	
Type:	single-seat tactical fighter
Powerplant:	two 1850kg (4080lb) thrust General Electric J85-GE-13 turbojets
Performance:	maximum speed 1487km/h (924mph) at 10,975m (36,000ft); service ceiling 15,390m (50,500ft); combat radius 314km (195 miles) with maximum warload
Weights:	empty 3667kg (8085lb); maximum take-off 9373kg (20,667lb)
Dimensions:	wing span 7.70m (25ft 3in); length 14.38m (47ft 2in); height 4.01m (13ft 2in); wing area 15.78m^2 (170 sq ft)
Armament:	two 20mm (0.79in) M39 cannon; up to 1995kg (4400lb) of stores on external pylons

Vought A-7 Corsair II

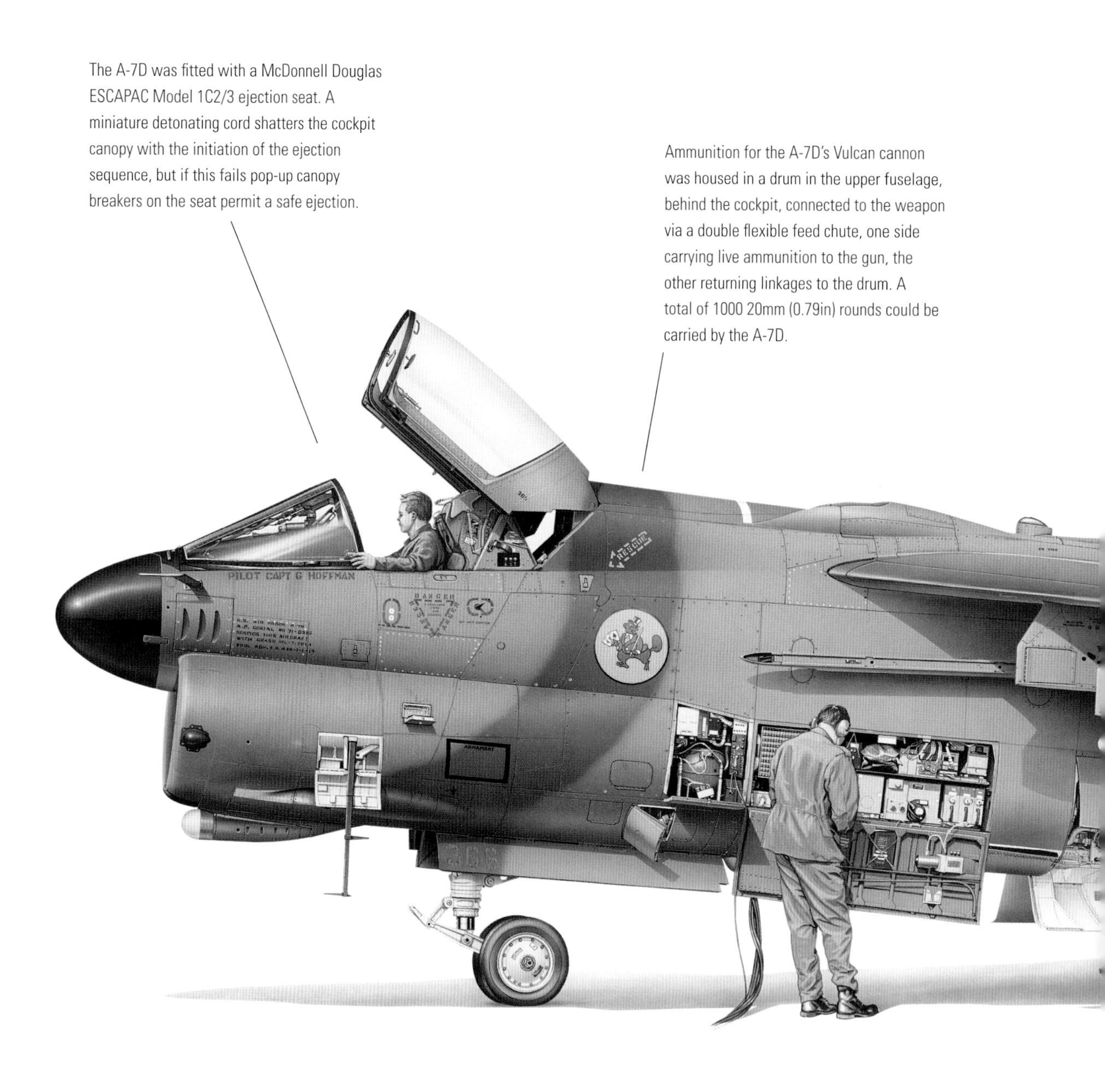

The A-7D pictured here was assigned to the 125th TFS, 138th TFG, of the Oklahoma Air National Guard, based at Tulsa. Before converting to the A-7D in July 1978, the Oklahoma ANG flew the F-100D/F Super Sabre, and began converting to the F-16C/D in 1993.

Mounted on LAU-7/A launch rails, seen here empty, the A-7E's pair of AIM-9L Sidewinder AAMs gave the aircraft a respectable self-defence capability. In fact, in manoeuvring combat at low level, a well-flown 'clean' A-7 could out-fight the majority of interceptors, thanks to its 7-g manoeuvre limit and 140 degrees per second roll rate.

On 11 February 1964 the US Navy announced that Ling-Temco-Vought was the winner of a design competition for a new single-seat carrier-based light attack aircraft to supplement and eventually replace the A-4E Skyhawk. The requirement, issued in May 1963, was for a subsonic aircraft capable of carrying a substantially greater load of non-nuclear weapons than the A-4E. Designated VAL (V for heavier-than-air, A for attack, L for lightweight), the projected aircraft was required to provide close support for ground forces up to 1125km (700 miles) inland from any coastline approachable by the US fleet. To keep costs to a minimum and ensure quick delivery, it was decided that the aircraft should be based on an existing design.

LTV's design study, based on the F-8E Crusader, was selected from more than two dozen proposals. An initial contract to develop and build three aircraft, under the designation A-7A, was awarded on 19 March 1964. It called for completion of a full-scale mock-up by 22 June 1964 and first flight in 1965. Although based on the design of the F-8 Crusader, the A-7 was in reality a completely different aircraft, being designed for high subsonic speed, a formula that made it possible to reduce structural weight dramatically, thereby increasing range and payload. Development was extremely rapid, losses sustained by the A-4 Skyhawk squadrons in Vietnam making it imperative to introduce the new type into combat as quickly as possible. The prototype flew for the first time on 27 September 1965, and several versions were subsequently produced for the US Navy and USAF by the Vought Corporation, a subsidiary of

Below: **The A-7E pictured here was assigned to Commander John Leenhouts, executive officer of Attack Squadron VA-72 and a leading exponent of naval Corsair operations. VA-72 flew from USS *John F. Kennedy* as part of CVW 3 for the duration of Operation Desert Storm.**

Ling-Temco-Vought. The first attack variant was the A-7A, which made its combat debut in the Gulf of Tonkin on 4 December 1966 with Attack Squadron VA-147, operating from USS *Ranger*. By the end of 1967 two more squadrons, VA-97 and VA-27, had formed from VA-122, the A-7 combat readiness training squadron for the West Coast, at NAS Lemoore, California, with another two, VA-82 and VA-86, forming from VA-174, the East Coast training squadron, at NAS Cecil Field, Florida. Also deployed to Southeast Asia was the A-7E, a close support/interdiction variant developed for the US Navy. By the end of the conflict in Vietnam, A-7s had flown more than 100,000 combat missions.

In all, 199 A-7As were delivered before production switched to the A-7B, which had an uprated engine. The first production model flew on 6 February 1968, the USN taking delivery of 198 examples. The next variant was the A-7D tactical fighter for the USAF, which went into action in Vietnam in October 1972; 459 were built, many being allocated to Air National Guard (ANG) units. The latter began taking delivery of the A-7D in October 1975, when the 188th Tactical Fighter Squadron at Kirtland AFB received its first aircraft. In total, 14 ANG squadrons operated the type and achieved an enviably low attrition rate. The final major Corsair variant, the two-seat A-7K, served only with the ANG. The last A-7D/K operators were

Above: **A-7 Corsair IIs of Navy Attack Squadron VA-46 Clansmen over the Gulf of Tonkin during the Vietnam War. By the end of the conflict in Vietnam, Corsairs had flown more than 100,000 combat missions. The USAF also used the A-7D Corsair in the tactical fighter role.**

OH

Left: **A-7D Corsair IIs of the Ohio Air National Guard, a major user of the type. Corsairs were deployed to the Middle East in 1990 and saw action during Operation Desert Storm in January/February 1991. None was lost to enemy action.**

the Ohio, Iowa and Oklahoma ANG units. From 1987 to 1988 48 A-7D and eight A-7K airframes were upgraded with the addition of a nose-mounted AN/AAR-49 FLIR for the Low-Altitude Navigation and Attack (LANA) system. LANA introduced an automatic terrain-following capability as well as a wide-angle HUD. The first LANA aircraft to be completed was an A-7K for the 150th Tactical Fighter Group, New Mexico ANG, introduced into service in the summer of 1987. Prior to LANA, navigation capability was limited to the nose-mounted Texas Instruments APQ-126(V) forward-looking radar (FLR), providing navigation and mapping pictures, limited terrain-following, terrain avoidance, air/ground ranging and weapon delivery information. A quarter of a century after it first saw combat in Vietnam, the A-7 was in action once more, attacking Iraqi targets during Operation Desert Storm in 1991. Based aboard the USS *John F. Kennedy* were VA-46 'Clansmen' and VA-72 'Bluehawks', whose A-7Es were tasked primarily with light attack and interdiction. The aircraft participated in the first air strikes, continuing to fly attack missions with HARM and Walleye air-to-surface missiles, as well as free-falling 'iron bombs', throughout the conflict. No A-7E was lost to enemy action during Desert Storm, although one aircraft was ditched after it sustained landing damage aboard the *Kennedy*.

Corsair IIs were also operated by the Hellenic, Portuguese and Thai air forces. Portugal's A-7P Corsair IIs, all ex-US Navy, were delivered from 1981 and withdrawn in 1997.

Specification: Vought A-7E Corsair II	
Type:	single-seat tactical fighter-bomber
Powerplant:	one 6802kg (15,000lb) thrust Allison TF41-A-2 turbofan
Performance:	maximum speed 1123km/h (698mph) at sea level; service ceiling 15,545m (51,000ft); range 1127km (700 miles)
Weights:	empty 8970kg (19,781lb); maximum take-off 19,047kg (42,000lb)
Dimensions:	wing span 11.81m (38ft 9in); length 14.04m (46ft 1in); height 4.90m (16ft 1in); wing area 34.84m² (375 sq ft)
Armament:	one 20mm (0.79in) M61 Vulcan cannon; provision for up to 6802kg (15,000lb) of external stores

General Dynamics F-111

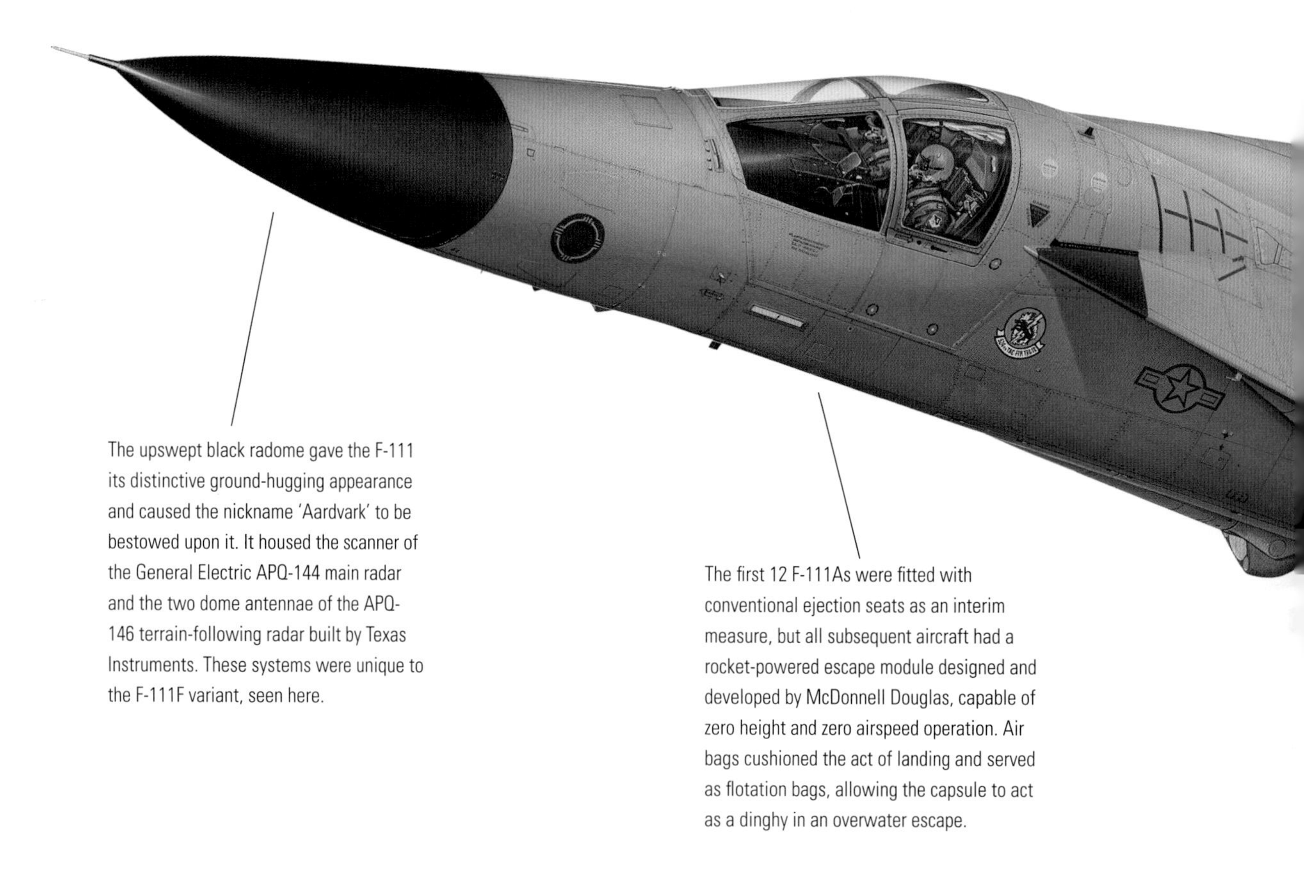

The upswept black radome gave the F-111 its distinctive ground-hugging appearance and caused the nickname 'Aardvark' to be bestowed upon it. It housed the scanner of the General Electric APQ-144 main radar and the two dome antennae of the APQ-146 terrain-following radar built by Texas Instruments. These systems were unique to the F-111F variant, seen here.

The first 12 F-111As were fitted with conventional ejection seats as an interim measure, but all subsequent aircraft had a rocket-powered escape module designed and developed by McDonnell Douglas, capable of zero height and zero airspeed operation. Air bags cushioned the act of landing and served as flotation bags, allowing the capsule to act as a dinghy in an overwater escape.

The aircraft pictured here is an F-111F of the 27th Fighter Wing, Cannon AFB, New Mexico. It formerly belonged to the 48th TFW at Lakenheath, Suffolk, England, and was reassigned to Cannon when the 48th re-armed with F-15Es. Cannon's last F-111 was retired in 1996, the type being replaced by F-16Cs.

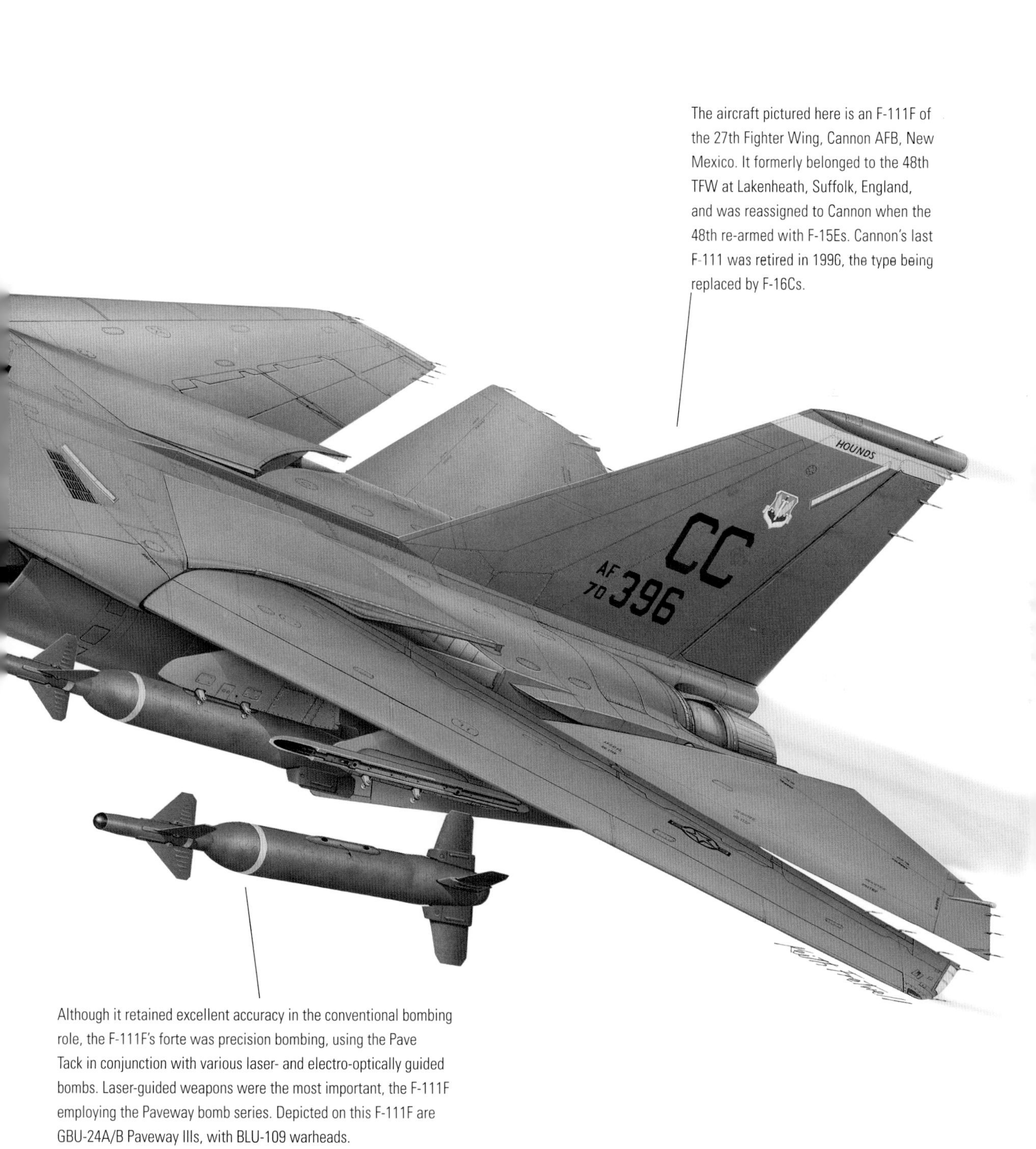

Although it retained excellent accuracy in the conventional bombing role, the F-111F's forte was precision bombing, using the Pave Tack in conjunction with various laser- and electro-optically guided bombs. Laser-guided weapons were the most important, the F-111F employing the Paveway bomb series. Depicted on this F-111F are GBU-24A/B Paveway IIIs, with BLU-109 warheads.

The Vietnam War saw the combat debut of an aircraft that has become synonymous with interdiction: the General Dynamics F-111. The development history of the F-111 began in 1962, when the General Dynamics Corporation, in association with Grumman Aircraft, was selected to develop a variable-geometry tactical fighter to meet the requirements of the USAF's TFX programme. An initial contract was placed for 23 development aircraft, including 18 F-111As for the USAF and five F-111Bs for the US Navy (in the event, the Navy cancelled the F-111 order). Powered by two Pratt & Whitney TF30-P-1 turbofan engines, the prototype F-111A flew for the first time on 21 December 1964, and during its second flight on 6 January 1965 the aircraft's wings were swept through the full range from 16 to 72.5 degrees.

One hundred and sixty production F-111As were built, the first examples entering service with the 4480th Tactical Fighter Wing at Nellis AFB, Nevada, in October 1967. On 17 March the following year six aircraft from this unit flew to Takhli AFB in Thailand for operational evaluation in Vietnam (Operation Combat Lancer), making their first

Below: **General Dynamics F-111As of the 27th Tactical Fighter Wing, Cannon AFB. The F-111A was tested under combat conditions in Vietnam, several being lost through structural failure, but when this problem was rectified the aircraft was a superb fighting machine.**

Above: **A General Dynamics F-111C of the Royal Australian Air Force, armed with Paveway laser-guided bombs. The acquisition of the F-111 by Australia was only secured after much political in-fighting, fuelled by uninformed press speculation about the aircraft's capabilities.**

sorties on 25 March. The operation ended unhappily when three of the aircraft were lost as a result of metal fatigue in a control rod, but the problem was rectified and in September 1972 the F-111As of the 429th and 430th Tactical Fighter Squadrons deployed to Takhli and performed very effective service in the closing air offensive of the war (Linebacker II), attacking targets in the Hanoi area at night and in all weathers through the heaviest anti-aircraft concentrations in the history of air warfare.

The F-111E variant, which superseded the F-111A in service, featured modified air intakes to improve performance above Mach 2.2. Re-equipment of the 20th TFW at Upper Heyford in the UK was completed in the summer of 1971 and the unit was assigned the war role of interdicting targets deep inside hostile territory as part of NATO's 2nd Allied Tactical Air Force. The other UK-based F-111 TFW was the 48th; based at Lakenheath in Suffolk, it was assigned to 4 ATAF in its war role and had the ability to interdict targets as far away as the Adriatic. The 48th TFW was armed with the F-111F, a fighter-bomber variant combining the best features of the F-111E and the FB-111A (the strategic bomber version) and fitted with the more powerful TF30-P-100 engines. The 48th TFW's aircraft were equipped to carry two B43 nuclear stores internally, as well as a variety of ordnance under six wing hardpoints, and formed the core of NATO's theatre nuclear strike force.

In the conventional role, the F-111F's primary precision attack weapon system was the Pave Tack self-contained pod containing a laser designator, rangefinder and forward-looking infrared (FLIR) equipment for use with laser-guided bombs like the 906kg (2000lb) Mk82 Snakeye, the GBU-15 TV-guided bomb or the Maverick TV-guided missile. The Pave Tack pod was stowed inside the F-111's weapons bay in a special cradle, rotating through 180 degrees to expose the sensor head when the system was activated. The sensor head provided the platform for the FLIR seeker, the laser designator and rangefinder, so that the Weapons System Operator (WSO) was presented with a stabilized infrared image, together with range information, on his display.

The F-111C (24 built) was a strike version for the RAAF, while the F-111D (96 built) was optimized for tactical support. Total production of the F-111, all variants, was 562 aircraft, including 23 development aircraft. The type saw considerable action subsequent to the Vietnam conflict, being employed in retaliatory strikes against Libya in April 1986 and in attacks on targets in Iraq during Operation Desert Storm in 1991.

Specification: General Dynamics F-111E

Type:	two-seat interdictor
Powerplant:	two 11,385kg (25,100lb) Pratt & Whitney TF-30-P-100 turbofans
Performance:	maximum speed 2655km/h (1650mph) at altitude; service ceiling 17,985m (59,000ft); range 4707km (2925 miles)
Weights:	empty 21,394kg (47,175lb); maximum take-off 45,359kg (100,000lb)
Dimensions:	wing span 19.20m (63ft) unswept, 9.74m (31ft 11in) swept; length 22.40m (73ft 6in); height 5.22m (17ft 1in); wing area 48.77m2 (525 sq ft) unswept
Armament:	one 20mm (0.79in) M61A-1 multi-barrelled cannon and one 340kg (750lb) B43 nuclear store, or two B43s in internal bay; provision for up to 14,290kg (31,000lb) of ordnance on eight underwing hardpoints

Fairchild Republic A-10 Thunderbolt II

The A-10 has standard daytime formation lights on the extreme tailcone, lower fins, wingtips, spine and belly. A modification known as LASTE (Low Altitude Safety and Target Enhancement) adds low-voltage lights to the wingtips, fins and spine just aft of the cockpit.

The massive GAU-8/A Avenger seven-barrel rotary cannon is driven by two hydraulic motors. It is spun up to its full firing rate of 4200 rounds per minute in 0.55 seconds, and has a maximum capacity of 1350 rounds of 30mm (1.18in) ammunition on a linkless feed system.

This aircraft carries an ALQ-184 ECM pod and two AIM-9L Sidewinders, the latter on an ANG-developed Dual Rail Adaptor (DRA).

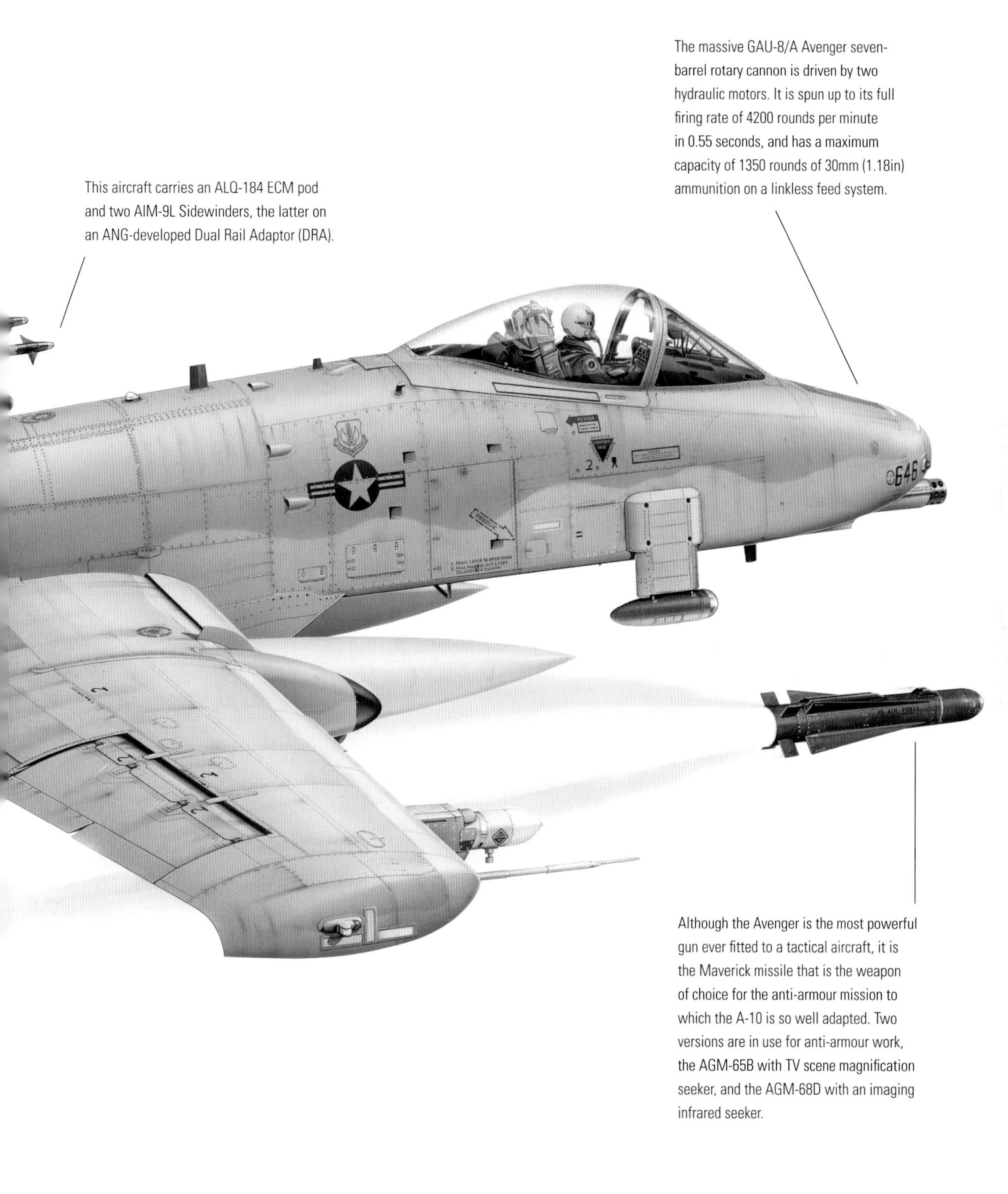

Although the Avenger is the most powerful gun ever fitted to a tactical aircraft, it is the Maverick missile that is the weapon of choice for the anti-armour mission to which the A-10 is so well adapted. Two versions are in use for anti-armour work, the AGM-65B with TV scene magnification seeker, and the AGM-68D with an imaging infrared seeker.

In December 1970 Fairchild Republic and Northrop were each selected to build a prototype of a new close support aircraft for evaluation under the USAF's A-X programme, and in January 1973 it was announced that Fairchild Republic's contender, the YA-10, had been chosen. Fairchild met the armour requirement by seating the pilot in what was almost a titanium 'bathtub', resistant to most firepower except a direct hit from a heavy-calibre shell, and added to this a so-called redundant structure policy whereby the pilot could retain control even if the aircraft lost large portions of its airframe, including one of the two rear-mounted engines. The core of the A-10's built-in firepower was its massive GAU-8/A seven-barrel 30mm (1.18in) rotary cannon, which was mounted on the centreline under the forward fuselage. The aircraft also had eight underwing and three under-fuselage attachments for up to 7250kg (16,000lb) of bombs, missiles, gun pods and jammer pods, and carried the Pave Penny laser system pod for target designation. It was fitted with very advanced avionics including a central air data computer, an inertial navigation system and a head-up display.

Left: **An A-10A Thunderbolt II undergoing trials with the 6512th Test Squadron at the USAF Flight Test Center, Edwards AFB. The black soot patches under the cockpit are caused by smoke streaming back from the rapid-firing GAU-8/A rotary cannon.**

The A-10 was designed to operate from short, unprepared strips less than 457m (1,500ft) long. Deliveries began in March 1977 to the 354th Tactical Fighter Wing at Myrtle Beach, South Carolina; in all, the USAF took delivery of 727 aircraft for service with its tactical fighter wings, the emphasis being on European operations. The A-10 had a combat radius of 463km (250 nautical miles), enough to reach a target area on the East German border from a Forward Operating Location (FOL) in central Germany and then move on to another target area in northern Germany. The aircraft had a three-and-a-half-hour loiter endurance, although operational war sorties in Europe

would probably have lasted between one and two hours. The operational tactics developed for the aircraft involved two A-10s giving one another mutual support, covering a swathe of ground two or three miles wide, so that an attack could be quickly mounted by the second aircraft once the first pilot had made his firing pass on the target. The optimum range for engaging a target was 1220m (4000ft), and the A-10's gunsight was calibrated for this distance. As the highly manoeuvrable A-10's turning circle was also 1220m (4000ft), this meant that the pilot could engage the target without having to pass over it. A one-second burst of fire would place seventy rounds of 30mm (1.18in) shells on the target and, as a complete 360-degree turn took no more than 16 seconds, a pair of A-10s could bring almost continuous fire to bear. The 30mm (1.18in) ammunition drum carried enough rounds to make ten to fifteen firing passes. In order to survive in a hostile environment dominated by radar-controlled AAA, A-10 pilots trained to fly at 30m (100ft) or lower, never remaining straight and level for more than four seconds. One of the aircraft's big advantages in approaching the combat zone was that its twin General Electric TF34-GE-100 turbofan engines were very quiet, so that it was able to achieve total surprise as it popped up over a contour of the land for weapons release. Attacks on targets covered by AAA involved close cooperation between the two A-10s; while one engaged the target the other stood off and engaged anti-aircraft installations with its TV-guided Maverick missiles, six of which were normally carried. The A-10 also had a considerable air-to-air capability, the tactic being for the pilot to turn towards an attacking fighter and use coarse rudder to spray it with 30mm (1.18in) shells.

In general, operations by the A-10s envisaged cooperation with US Army helicopters; the latter would hit the mobile SAM and AAA systems accompanying a Soviet armoured thrust and, with the enemy's defences at least temporarily stunned or degraded, the A-10s would be free to concentrate their fire on the tanks. Twelve years later, these tactics were used to deadly effect in the 1991 Gulf War. In that conflict, the A-10 went into action against much the same type of equipment that it would have been likely to meet in action against Warsaw Pact armoured forces in northern and central Europe, namely the T-70 main battle tank, tracked anti-aircraft vehicles and armoured personnel carriers. It destroyed them all with an efficiency that was almost clinical.

Above: **Designed to counter strong enemy armoured thrusts in northern Europe, the A-10A showed what it could do during the 1991 Gulf War, when its firepower was brought to bear on the latest Russian-designed armour used by Iraq.**

Specification: Fairchild Republic A-10A Thunderbolt II	
Type:	single-seat close support and assault aircraft
Powerplant:	two 4111kg (9065lb) thrust General Electric TF34-GE-100 turbofans
Performance:	maximum speed 706km/h (439mph) at sea level; service ceiling 7625m (25,000ft); combat radius 463km (250 nautical miles) with two-hour loiter
Weights:	empty 11,321kg (24,963lb); maximum take-off 22,680kg (50,000lb)
Dimensions:	wing span 17.53m (57ft 6in); length 16.26m (53ft 4in); height 4.47m (14ft 8in); wing area 47.01m2 (506 sq ft)
Armament:	one 30mm (1.18in) GAU-8/A rotary cannon with 1350 rounds; 11 hardpoints for up to 7556kg (16,000lb) of ordnance, including Rockeye cluster bombs, Maverick ASMs and SUU-23 20mm (0.79in) cannon pods

Grumman F-14A Tomcat

The rear cockpit of the F-14 contains the detailed data display (DDD), on which targets are displayed in range rate-versus-azimuth format for pulse-Doppler operation, and in range-versus-azimuth for pulse operation.

This F-14A served with VF-111 'Sundowners' and was deployed aboard the USS *Carl Vinson* (CVN-70) with the US Pacific Fleet in the 1980s. The Tomcat wears a low-visibility overall light grey camouflage scheme. On the nose are suitably toned-down 'sharkmouth' markings, also repeated on the external fuel tanks.

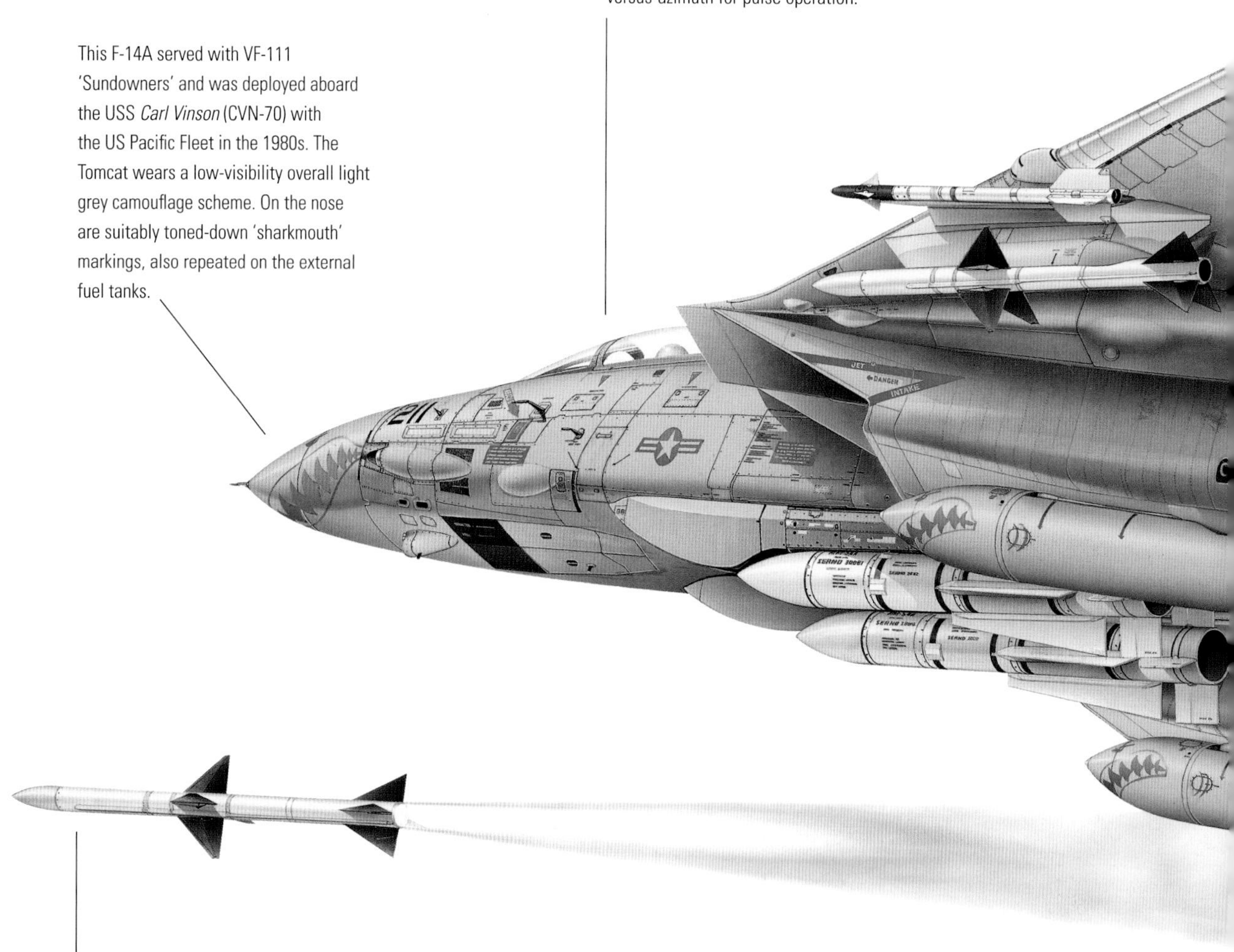

Raytheon AIM-7 Sparrow missiles are the F-14's primary medium-range (beyond visual range) weapons. Carried originally in AIM-7F but now in improved AIM-7M form, the missile is guided by a semi-active radar homing head which tracks the fighter's own radar signals reflected from the target.

Several problems were encountered during the introduction of the Tomcat into Fleet service. Poor logistics forced squadrons to cannibalize many aircraft in order to keep others airworthy. In the air, buffeting was experienced when the flaps were deployed, and fatigue cracking affected the boat-tail fairing; both problems were easily cured.

F-14D
NAVY
161867

Although its development history was beset by problems, the variable-geometry F-14 Tomcat emerged from them all to become one of the most formidable interceptors of all time, designed from the outset to establish complete air superiority in the vicinity of a carrier task force and also to attack tactical objectives as a secondary role. Selected in January 1969 as the winner of a US Navy contest for a carrier-borne fighter (VFX) to replace the Phantom, the prototype F-14A flew for the first time on 21 December 1970 and was followed by 11 development aircraft. The variable-geometry fighter completed carrier trials in the summer of 1972 and deliveries to the US Navy began in October that year, the Tomcat forming the interceptor element of a carrier air wing. At the heart of the Tomcat's offensive capability is the Hughes AN/AWG-9 weapons control system, which enables the two-man crew to detect airborne targets at ranges of up to 315km (170nm), depending on their size, and cruise missiles at 120km (65nm). The system can track 24 targets and initiate attacks on six of them at the same time, at a variety of altitudes and ranges. The Tomcat's built-in armament consists of one General Electric M61A-1 Vulcan 20mm (0.79in) gun mounted in the port side of the forward fuselage, with 675 rounds. Main missile armament comprises four Sparrow AAMs partially recessed under the fuselage, or four Phoenix AAMs mounted below the fuselage. In addition, four Sidewinder AAMs, or two Sidewinders plus two Phoenix or two Sparrows, can be carried on underwing pylons. The Tomcat can carry a mixture of ordnance up to a maximum of 6576kg (14,500lb), and is fitted with a variety of ECM equipment.

A task force's Tomcats are normally required to fly three kinds of mission: Barrier Combat Air Patrol, Task Force CAP and Target CAP. Barrier CAP involves putting up a defensive screen at a considerable distance from the task force under the direction of a command and control aircraft. Since fighters flying Barrier CAP are likely to encounter the greatest number of incoming enemy aircraft, Tomcats usually carry their full armament of six Phoenix AAMs. These weapons, which carry a 60kg (132lb) warhead, reach a speed of more than Mach 5 and have a range of over 200km (125 miles), which makes them highly suitable for long-range interception of aircraft flying at all levels and also sea-skimming missiles. Hostile aircraft that survive the attentions of the Tomcats on Barrier CAP are engaged by fighters of the Task Force CAP, which operate within sight of

Left: **The F-14D Tomcat, pictured here, is an improved version with more powerful radar, enhanced avionics, a redesigned cockpit and a tactical jamming system. Thirty-seven aircraft were built from new and 18 converted from F-14As.**

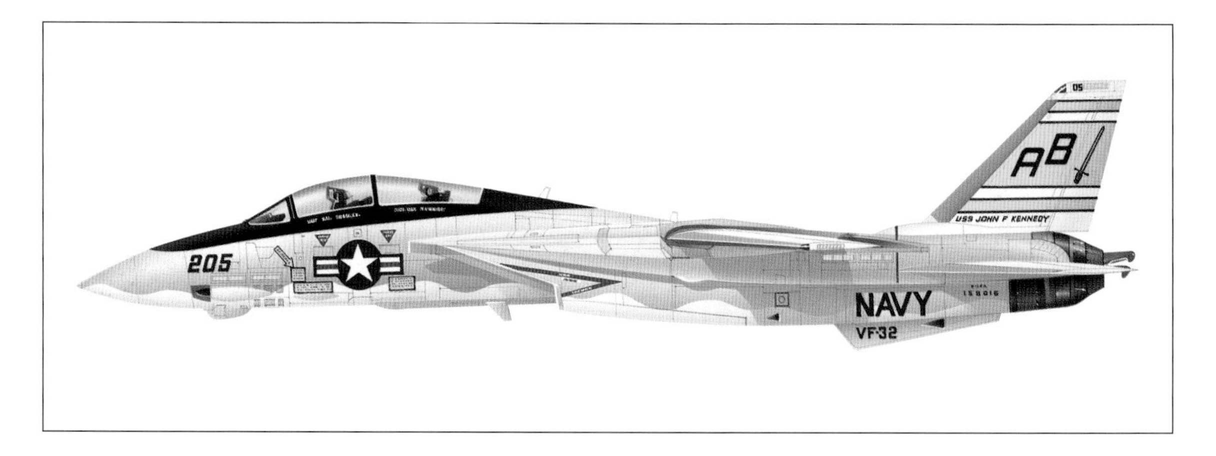

Above: **A Grumman F-14A Tomcat of Navy Fighter Squadron VF-32, operating from the aircraft carrier USS *John F. Kennedy*. The Tomcat has a formidable combat air patrol capability, with its ability to intercept hostile aircraft far out from a task force.**

Right: **An F-14D Tomcat of US Navy fighter squadron VF-31. The F-14D, which first flew in February 1990, is a much improved version with digital avionics, digital radar processing and displays, improved ejection seats and a redesigned cockpit.**

the ships and which are armed with a mixture of Phoenix, Sparrow and Sidewinder AAMs. If targets still show signs of breaking through and all defensive AAMs are expended, the Tomcats can continue the engagement with their Vulcan cannon at close range. The Tomcat's two Pratt & Whitney TF30-P-414 turbofans give it a maximum low-level speed of 910mph (Mach 1.2) and a high-level speed of 1544mph, or Mach 2.34.

Development of the production F-14A was hampered by the loss of the prototype in December 1970, but 478 aircraft were supplied to the US Navy in total, and 80 more F-14As were exported to Iran in the later 1970s. The F-14B, a proposed version with Pratt & Whitney F401-PW-400 turbofans, was cancelled, but 32 F-14As were fitted with the General Electric F110-GE-400 and redesignated F-14B. The F-14D is an improved version with more powerful radar, enhanced avionics, a redesigned cockpit and a tactical jamming system; 37 aircraft were built from new and 18 converted from F-14As.

In the 1980s the Tomcat was involved in several clashes with Libyan fighters over the Gulf of Sirte. On 19 August 1981, for example, two aircraft of VF-41 were scrambled from the USS *Nimitz* to investigate two radar contacts approaching an area where the US Sixth Fleet was holding an exercise. The two radar returns continued to close head-on with the Tomcats, and the weapons officers of both American aircraft reported that their equipment had locked on to transmissions from Soviet-made SRD-5M 'High Fix' air interception and fire control radars operating in the I-band. This type of radar was known to be installed in the intake centrebody of the Sukhoi Su-22 Fitter, a type in service with the Libyan Air Force. Both Fitters were destroyed by Sidewinder AAMs during the ensuing engagement, the Libyan aircraft having launched AA-2 Atoll infrared homing missiles at the F-14s.

During the Gulf War of 1991 the type shared the air combat patrol task with the McDonnell Douglas F-15 Eagle. Since then the type has seen active service in the Balkans and Afghanistan, and has been involved in enforcing the 'No-Fly Zones' in Iraq.

Specification: Grumman F-14A Tomcat

Type:	two-seat fleet defence interceptor
Powerplant:	two 9480kg (20,900lb) thrust Pratt & Whitney TF30-P-412A turbofans
Performance:	maximum speed 2517km/h (1564mph) at altitude; service ceiling: 17,070m (56,000ft); range 1994km (1239 miles) with full weapons load
Weights:	empty 18,190kg (40,104lb); maximum take-off 33,717kg (74,349lb)
Dimensions:	wing span 19.45m (63ft 8in) unswept, 11.65m (38ft 2in) swept; length 19.10m (62ft 8in); height 4.88m (16ft 0in); wing area 52.49m^2 (565 sq ft)
Armament:	one 20mm (0.79in) M61A-1 Vulcan rotary cannon, plus a combination of AIM-7 Sparrow medium range AAMs, AIM-9 short range AAMs, and AIM-54 Phoenix long range AAMs

200
NAVY
VF-31
NK
CAG

Lockheed-Martin F-16

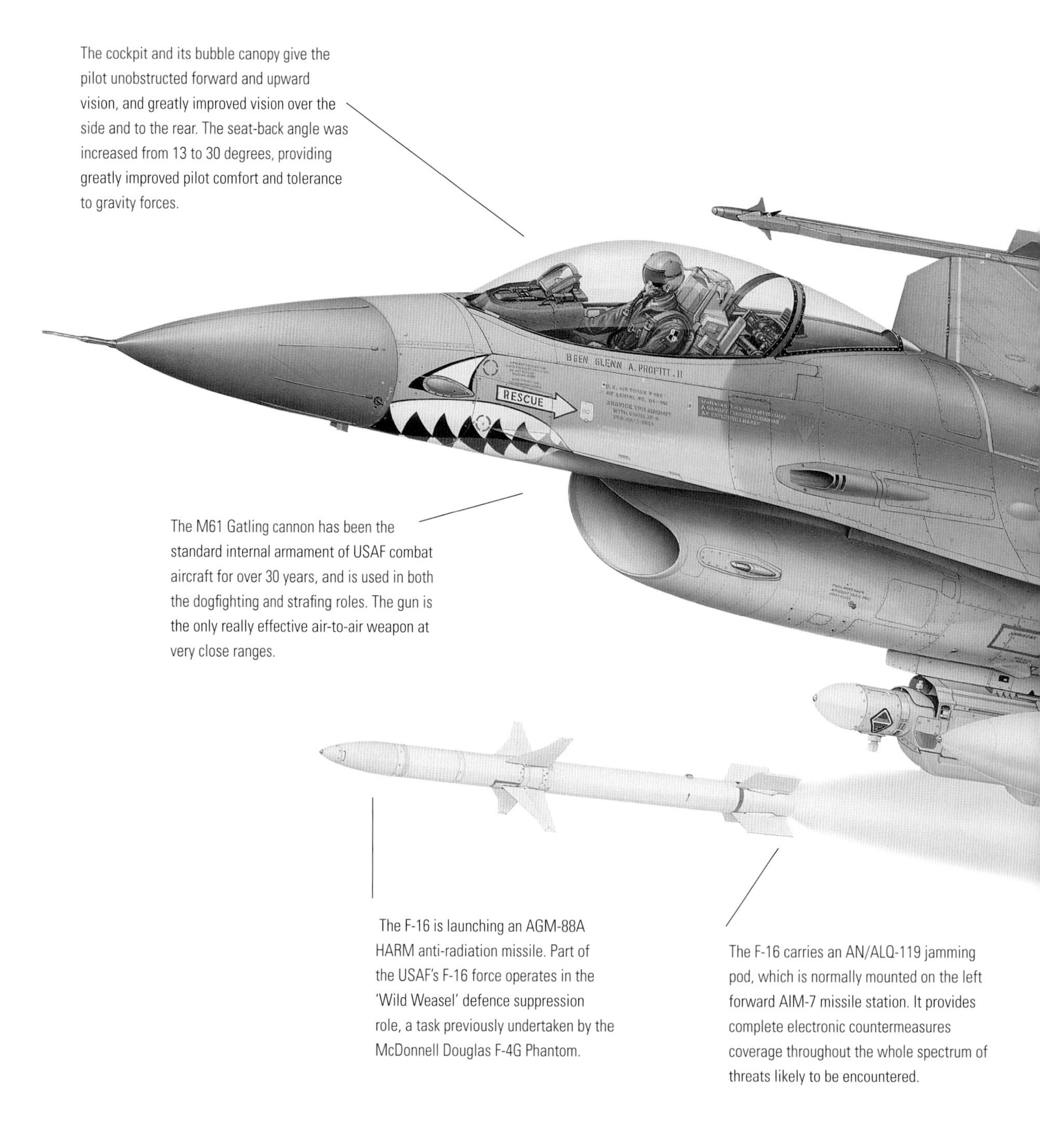

The cockpit and its bubble canopy give the pilot unobstructed forward and upward vision, and greatly improved vision over the side and to the rear. The seat-back angle was increased from 13 to 30 degrees, providing greatly improved pilot comfort and tolerance to gravity forces.

The M61 Gatling cannon has been the standard internal armament of USAF combat aircraft for over 30 years, and is used in both the dogfighting and strafing roles. The gun is the only really effective air-to-air weapon at very close ranges.

The F-16 is launching an AGM-88A HARM anti-radiation missile. Part of the USAF's F-16 force operates in the 'Wild Weasel' defence suppression role, a task previously undertaken by the McDonnell Douglas F-4G Phantom.

The F-16 carries an AN/ALQ-119 jamming pod, which is normally mounted on the left forward AIM-7 missile station. It provides complete electronic countermeasures coverage throughout the whole spectrum of threats likely to be encountered.

The tail markings of this F-16C denote that the aircraft belongs to the 52nd Tactical Fighter Wing, based at Spangdahlem in Germany. The 52nd has been one of NATO's front-line units for many years, and was formerly equipped with F-4 Phantoms.

The Lockheed Martin F-16 Fighting Falcon is the world's most prolific combat aircraft, with over 2000 in service with the United States Air Force and a further 2000 in service with 19 other air forces around the world. Current orders (in 2002) include Bahrein (10), Greece (50), Egypt (24), New Zealand (28), United Arab Emirates (80), Singapore (20), South Korea (20), Oman (12) and Chile (10). Israel, with the world's largest F-16 fleet outside the United States, plans to procure 110 F-16I aircraft, deliveries beginning in 2003. These aircraft will be fitted with Pratt & Whitney F100-PW-229 engines, Elbit avionics, Elisra electronic warfare systems and Rafael weapons and sensors, including Litening II laser target designating pods. Italy is to lease 34 aircraft until Eurofighter Typhoon is deployed, and Hungary is to acquire 24 ex-USAF aircraft.

The F-16, designed and built by General Dynamics, had its origin in a USAF requirement of 1972 for a lightweight fighter and first flew on 2 February 1974. It carries an advanced GEC-Marconi HUDWACS (HUD and Weapon Aiming Computer System) in which target designation cues are shown on the head-up display as well as flight symbols. The HUDWAC computer is used to direct the weapons to the target, as designated on the HUD. The F-16 HUDWAC shows horizontal and vertical speed, altitude, heading, climb and roll bars and range-to-go information for flight reference. There are five ground attack modes and four air combat modes. In air combat, the 'snapshoot' mode lets the pilot aim at crossing targets by drawing a continuously computed impact line (CCIL) on the HUD. The lead-computing off sight (LCOS) mode follows a designated target; the dogfight mode combines snapshoot and LCOS; and there is also an air-to-air missile mode. The F-16's underwing hardpoints are stressed for manoeuvres up to 9g, enabling the F-16 to dogfight while still carrying weaponry. The F-16B and D are two-seat versions, while the F-16C, delivered from 1988, featured numerous improvements in avionics and was available with a choice of engine. F-16s have seen action in the Lebanon (with the Israeli Air Force), in the Gulf War and the Balkans. A typical stores load might include two wingtip-mounted Sidewinders, with four more on the outer underwing stations; a podded GPU-5/A 30mm cannon on the centreline; drop tanks on the inboard underwing and fuselage stations; a Pave Penny laser spot tracker pod along the starboard side of the nacelle; and bombs, ASMs and flare pods on the four inner underwing stations. The aircraft can carry advanced beyond-visual-range missiles, Maverick ASMs, HARM and Shrike anti-radar missiles, and a weapons dispenser carrying various types of sub-munition including runway denial bombs, shaped-charge bomblets, anti-tank and area denial mines.

The F-16 has been constantly upgraded to extend its life well into the 21st century. The latest upgrade involves 650 Block 40/50 USAF F-16s, which are being updated under the Common Configuration Implementation Program (CCIP). The first aircraft was completed in January 2002, and the first phase programme provides core computer and colour cockpit modifications; the second phase, begun in September 2002, involves fitting the advanced interrogator/responder and Lockheed Martin Sniper XR advanced FLIR (Forward-Looking Infra-Red) targeting pod; while the

Below: **Unlike the majority of F-16 operators, Greece dispensed with the traditional USAF camouflage scheme, opting instead for a complex two-tone blue and grey scheme, as seen in this illustration of a Paveway-armed aircraft.**

Above: **The F-16 has been constantly upgraded to extend its life well into the 21st century. The first aircraft to undergo the latest upgrade was completed in January 2002. The programme involves several phases, each one involving different aspects of the F-16's equipment.**

third phase, starting in July 2003, adds Link 16 datalink, the Joint Helmet-Mounted Cueing System and an electronic horizontal situation indicator. The export version of the Sniper XR pod, the Pantera, has been selected by the Royal Norwegian Air Force. The Sniper XR (Extended Range) incorporates a high-resolution mid-wave FLIR, dual-mode laser, CCD TV, laser spot tracker and laser marker, combined with advanced image processing algorithms. A version of the Fighting Falcon for the US Navy, the F-16N, was ordered in the mid-1980s, and featured a strengthened wing capable of mounting equipment such as an Air Combat Manoeuvring Instrumentation (ACMI) pod. The first F-16N flew on 24 March 1987 and 26 aircraft were delivered, including four TF-16N two-seat trainers. Most were used by the US Navy's Top Gun fighter schools.

Specification: Lockheed-Martin F-16C

Type:	Single-seat air superiority and strike fighter
Powerplant:	either one 10,800kg (23,810lb) Pratt & Whitney F100-PW-200 or one 13,150kg (28,984lb) thrust General Electric F110-GE-100 turbofan
Performance:	maximum speed 2142km/h (1320mph) at altitude; service ceiling 15,240m (50,000ft); range 925km (575 miles)
Weight:	empty 16,057kg (35,400lb)
Dimensions:	wing span 9.45m (31ft); length 15.09m (49ft 6in); height 5.09m (16ft 8in); wing area 27.87m² (300 sq ft)
Armament:	one General Electric M61A1 multi-barrelled cannon; seven external hardpoints for up to 9276kg (20,450lb) of ordnance

McDonnell Douglas F/A-18 Hornet

The wingtip hardpoint is normally reversed to carry a version of the Sidewinder AAMs. Seen here is the AIM-9M, which has now been joined by the advanced AIM-9X.

Mounted in the nose is a General Electric M61A1 Vulcan 20mm (0.79in) rotary cannon provided with 570 rounds of ammunition, and firing at a rate of 6000 rounds per minute.

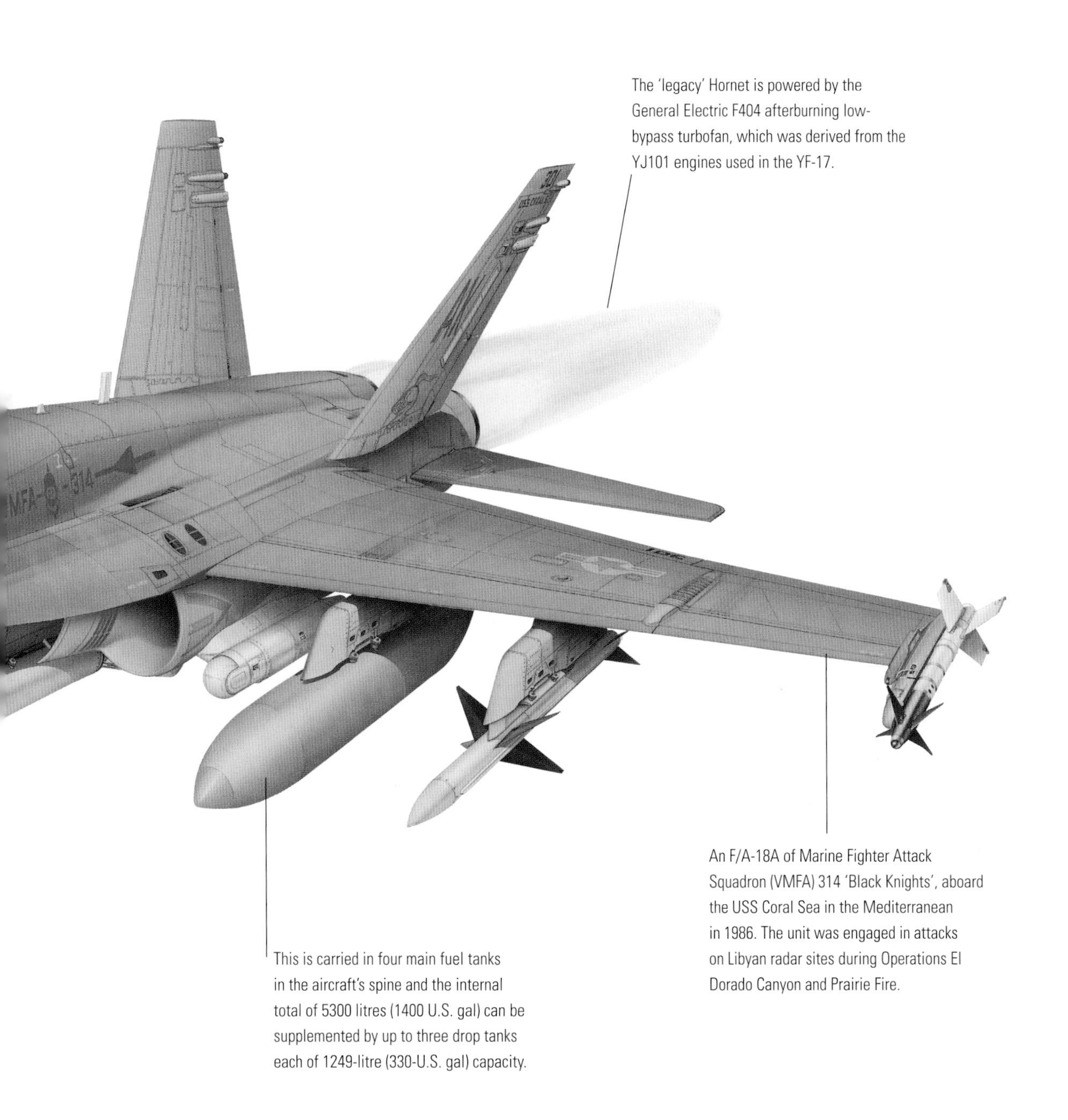

The 'legacy' Hornet is powered by the General Electric F404 afterburning low-bypass turbofan, which was derived from the YJ101 engines used in the YF-17.

An F/A-18A of Marine Fighter Attack Squadron (VMFA) 314 'Black Knights', aboard the USS Coral Sea in the Mediterranean in 1986. The unit was engaged in attacks on Libyan radar sites during Operations El Dorado Canyon and Prairie Fire.

This is carried in four main fuel tanks in the aircraft's spine and the internal total of 5300 litres (1400 U.S. gal) can be supplemented by up to three drop tanks each of 1249-litre (330-U.S. gal) capacity.

Replacing the F-4 Phantom II in the fleet air defence role, and the A-7 Corsair II attack aircraft, the Hornet brought the U.S. Navy's carrier air wing into a new era with a genuine multi-role carrier fighter, and the 'legacy' F/A-18 also continues to serve with the U.S. Marine Corps and a number of export operators.

The U.S. Navy originally selected the McDonnell Douglas Hornet under the Air Combat Fighter programme of the mid-1970s, in which the Northrop YF-17 emerged victorious against the General Dynamics YF-16. After the land-based YF-17 had been further developed and navalized, the result was the F/A-18, which would combine fighter and strike/attack roles in a single airframe, thanks to the use of a versatile airframe and advanced mission software. Northrop, as design originator, was to have responsibility for future land-based versions of the Hornet,

Opposite: **An F/A-18D of the U.S. Marine Corps' VMFA(AW)-225 armed with 127mm (5in) Zuni unguided rockets underwing, used for target-marking purposes in the forward air controller (airborne) role.**

while McDonnell Douglas would take design leadership and a majority workshare on the naval version for the U.S. Navy and Marine Corps. In the event, it was the naval model that also won export orders, although the two companies continued to share the responsibility for development and production. Today, McDonnell Douglas is part of Boeing, while Northrop has become Northrop Grumman. These two firms are now responsible for the F/A-18E/F Super Hornet, an entirely reworked aircraft, and its EA-18G Growler electronic attack derivative.

The first prototype of what is now referred to as the 'legacy' Hornet (to distinguish it from the Super Hornet) took to the air in November 1978 and was part of a batch of 11 development aircraft, two of which were completed as two-seaters. Production of the initial single-seat F/A-18A yielded 371 aircraft, the first of which were delivered in May 1980. The first fully operational unit was the Marine Corps' squadron VMFA-314, declared operational in January 1983. A first U.S. Navy unit followed suit in August 1983.

The Hornet first saw combat during the 1986 raids on Libya, flying from the carrier USS Coral Sea. Since then, the Hornet has been at the heart of U.S. Navy carrier and Marine Corps combat operations.

The F/A-18A gave way to the F/A-18C that became the dominant 'legacy' production variant, a total of 347 being ordered for U.S. Navy and Marine Corps service. First flown in September 1986, the C-model is equipped with the AN/APG-73 synthetic-aperture ground-mapping radar compatible with the AIM-120 AMRAAM and includes enhanced self-protection systems. Other changes include a new NACES ejector seat and small strakes above the wing leading edge root extensions. A night attack capability was added from the 138th F/A-18C, including night vision goggles, full-colour displays, a moving-map function and the AN/AAS-38 forward-looking infra-red pod. The night attack F/A-18C version began to be delivered to operational units in November 1989.

A two-seat version of the Hornet was developed alongside the F/A-18A, and produced the F/A-18B, which, apart from the addition of a second crew position, was basically unchanged from the A-model. A total of 40 aircraft were completed for the U.S. Navy and Marine Corps.

Left: **An F/A-18C aircraft assigned to Strike Fighter Squadron (VFA) 136 'Knighthawks' unloads a flare over the Persian Gulf before heading into Afghanistan for a close air support mission. As of October 2013, the U.S. Navy and Marine Corps included a total of 620 F/A-18A/B/C/D Hornets in operational and reserve service and in test roles.**

The F/A-18B has not been employed in an operational role in contrast to the definitive 'legacy' two-seater, the F/A-18D. This is essentially similar to the F/A-18C and after 31 production aircraft had been completed, the D-model also switched to a night attack version, production concluding with delivery of the 109th example. The night attack F/A-18D is primarily employed by the U.S. Marine Corps, with which it replaced the A-6 Intruder.

Land-based Hornets have been acquired by seven nations, and the first export customer for the type was Canada, which took delivery of 98 single-seat CF-188As and 40 two-seat CF-188Bs between 1982 and 1988. Australia was next, taking 57 F/A-18As and 18 F/A-18Bs. Spain acquired 60 EF-18As and 12 EF-18Bs. Thereafter, export aircraft were all completed to F/A-18C/D standard, comprising 32 F/A-18Cs and eight F/A-18Ds for Kuwait; 26 F/A-18Cs and eight F/A-18Ds for Switzerland; 57 F-18Cs and seven F-18Ds (initially without attack capability) for Finland and eight F/A-18Ds for Malaysia.

Upgrade of the 'legacy' Hornet was spearheaded by the U.S. Navy, and most export operators followed suit. The AN/ASQ-228 Advanced Targeting Forward-Looking Infra-Red (ATFLIR) pod has replaced the AN/AAS-38 and the range of air-to-ground ordnance has been expanded. The U.S. Navy modified 61 F/A-18As to the A+ configuration, with AN/APG-73 radar and F/A-18C avionics. Of the A+ conversions, 54 have been upgraded to full F/A-18C capability (as F/A-18A++). The F/A-18C+ configuration includes Link 16 datalink, colour cockpit displays, a moving-map display, AN/ALE-47 infra-red countermeasures, NACES and the Joint Helmet-Mounted Cueing System.

Specification: McDonnell Douglas F/A-18 Hornet

Type:	carrier-based fighter and strike/attack
Powerplant:	2 x 78.73kN (17,700lb) General Electric F404-GE-402 afterburning turbofans
Performance:	maximum speed more than 1915km/h (1190mph) at high altitude; service ceiling 115,240m (50,000ft); range 3336km (2073 miles)
Weights:	21,888kg (48,253lb) maximum
Dimensions:	wing span 11.43m (37ft 6in) without wingtip missiles; length 17.07m (56ft 0in); height 6.46m (21ft 2in); wing area 61.45m^2 (661.5 sq ft)
Armament:	1 x 20mm (0.79in) rotary cannon and up to 6200kg (13,700lb) of disposable stores

McDonnell Douglas F-15 Eagle

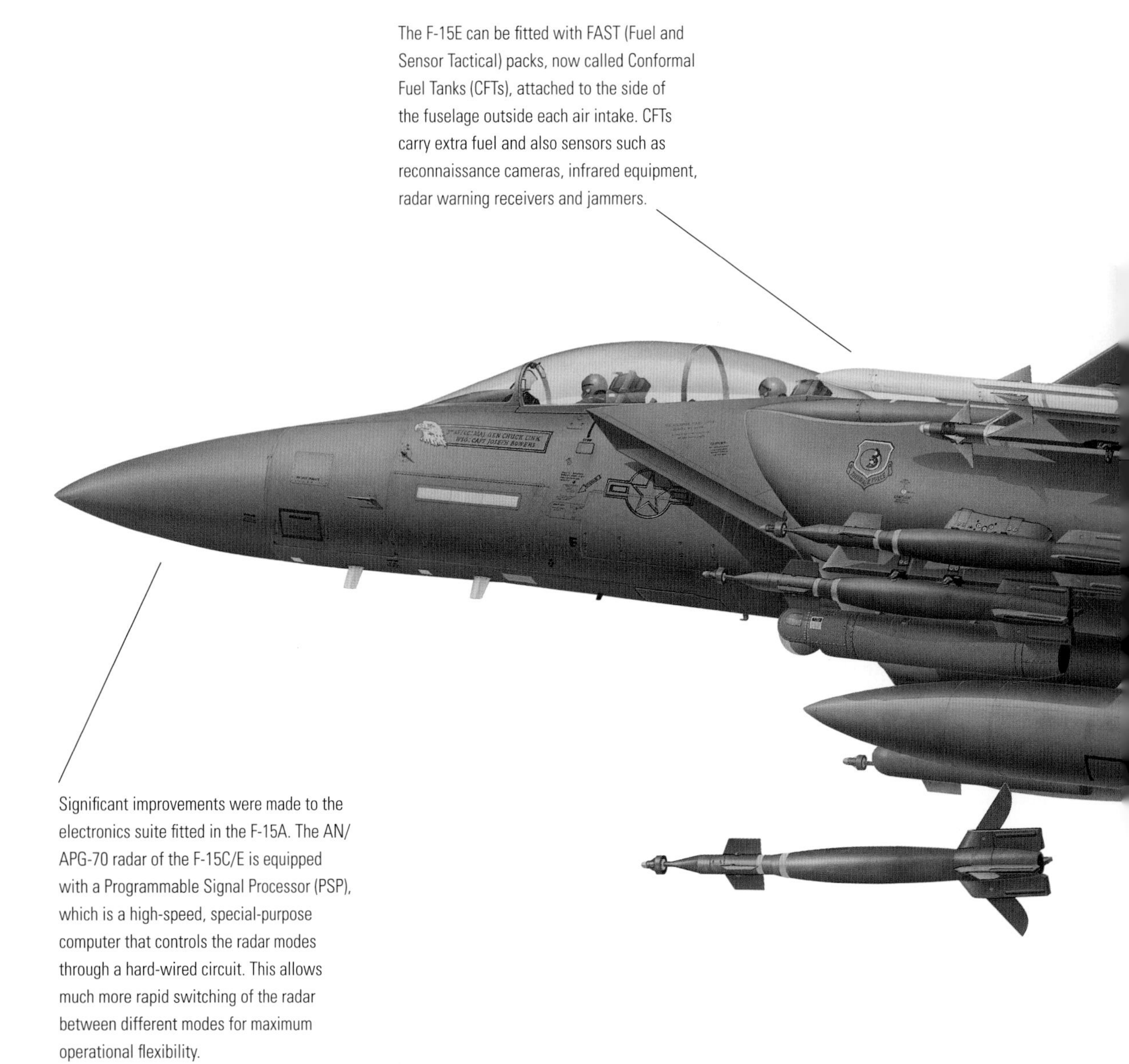

The F-15E can be fitted with FAST (Fuel and Sensor Tactical) packs, now called Conformal Fuel Tanks (CFTs), attached to the side of the fuselage outside each air intake. CFTs carry extra fuel and also sensors such as reconnaissance cameras, infrared equipment, radar warning receivers and jammers.

Significant improvements were made to the electronics suite fitted in the F-15A. The AN/APG-70 radar of the F-15C/E is equipped with a Programmable Signal Processor (PSP), which is a high-speed, special-purpose computer that controls the radar modes through a hard-wired circuit. This allows much more rapid switching of the radar between different modes for maximum operational flexibility.

The F-15E Strike Eagle pictured here bears the markings of the 48th Fighter Wing, based at Lakenheath (code letters LN) in the UK. Formerly equipped with F-111Fs, the 48th FW has deployed its F-15Es to both the Middle East and the former Yugoslavia as part of multinational peacekeeping forces.

A feature of the F-15 is its large wing area and correspondingly low wing loading. While this makes for a more uncomfortable low-level ride than that experienced by the crews of dedicated attack aircraft like the Su-24 or Tornado, it confers superb air-to-air capability on the F-15E, enabling it to put up a highly credible defence if attacked.

The United States Air Force and various aircraft companies in the USA began discussions on the feasibility of just such an aircraft and its associated systems to replace the F-4 Phantom in 1965, and four years later it was announced that McDonnell Douglas had been selected as prime airframe contractor for the new aircraft, then designated FX; as the F-15A Eagle, it flew for the first time on 27 July 1972, and first deliveries of operational aircraft were made to the USAF in 1975. Urgent impetus was given to the programme following the appearance of the Soviet MiG-25 Foxbat interceptor, an aircraft that was itself developed to meet a potential threat from a new generation of US strategic bombers like the North American XB-70 Valkyrie, a project that was subsequently cancelled.

The tandem-seat F-15B was developed alongside the F-15A, and the main production version was the F-15C. The latter was built under licence in Japan as the F-15J. The F-15E was supplied to Israel as the F-15I and to Saudi Arabia as the F-15S. The F-15E Strike Eagle is a dedicated strike/attack variant and, while the F-15C established and maintained air superiority, was at the forefront of precision bombing operations in the 1991 Gulf War. The F-15E was supplied to Israel as the F-15I and to Saudi Arabia as the F-15S. In all, the USAF took delivery of 1286 F-15s (all versions), Japan 171, Saudi Arabia 98 and Israel 56. F-15s saw much action in the 1991 Gulf War, and Israeli aircraft were in combat with the Syrian Air Force over the Bekaa Valley in the 1980s.

Simply stated, the F-15 Eagle was designed to out-perform, out-fly and out-fight any opponent it might encounter in the foreseeable future, in engagements extending from beyond visual range (BVR) right down to close-in turning combat. To fulfil this mission, the aircraft's design incorporated many innovations. The F-15's wing, for example, was given a conical camber and an airfoil section optimized to reduce wave drag at high speed. The last 20 per cent of the chord was thickened to delay boundary layer separation and so reduce drag. Manoeuvre performance is also enhanced by the slab tailplanes, which operate differentially, in concert with the ailerons, in the rolling plane and together for pitch control. They compensate to a great extent for loss of aileron effectiveness at extreme angles of attack, a vital factor in tight air combat. The aircraft's twin fins are also positioned to receive vortex flow off the wing in order to maintain directional stability at high angles of attack.

The F-15C – the main interceptor version – has a wing loading of only 25kg/m^2 (54 lb/sq ft) and this, together with two 10,779kg (23,770lb) thrust Pratt & Whitney F100-PW-220 advanced technology turbofans, gives it an

Below: **The F-15E Strike Eagle is a development of the basic F-15 design, combining unsurpassed air-to-air capability and superior air-to-ground capability in a single aircraft. The F-15E first went into action during operation Desert Storm in 1991.**

Above: **Seen in the markings it carried in the early 1990s, this F-15C of the 33rd Fighter Wing accounted for four of the 58th Tactical Fighter Squadron's 16 kills during the Gulf War. Among its victorious pilots were Colonel Rick Parsons and Captain Anthony R. Murphy.**

extraordinary turning ability and the combat thrust-to-weight ratio (1.3:1) necessary to retain the initiative in a fight. The high thrust-to-weight ratio permits a scramble time of only six seconds, using 183m (600ft) of runway, and a maximum speed of more than Mach 2.5 gives the pilot the margin he needs if he has to break off an engagement.

To increase the Eagle's survivability, redundancy is incorporated in its structure; for example, one vertical fin, or one of three wing spars, can be severed without causing the loss of the aircraft. Redundancy is also inherent in the F-15's twin engines, and its fuel system incorporates self-sealing features and foam to inhibit fires and explosions.

Primary armament of the F-15 is the AIM-7F Sparrow radar-guided AAM, with a range of up to 35 miles. The Eagle carries four of these, backed up by four AIM-9L Sidewinders for shorter range interceptions and a General Electric 20mm (0.79in) M61 rotating barrel cannon for close-in combat. The gun is mounted in the starboard wing root and is fed by a fuselage-mounted drum containing 940 rounds. The aircraft's Hughes AN/APG-70 pulse-Doppler air-to-air radar provides a good look-down capability and can be used in a variety of modes; it can pick up targets at around 185 km (100 nm) and, in the raid assessment mode, can resolve close formations into individual targets, giving the F-15 pilot an important tactical advantage.

When the radar detects a target in the basic search mode the pilot directs the AN/APG-70 to lock on and track by putting a bracket over the radar return, using a selector mounted on the control column. The locked-on radar will then show attack information such as target closing speed, range, bearing, altitude separation and parameters governing the F-15's weapons release. When the target enters the kill envelope of the weapon selected, the pilot decides whether to attack using his head-down, virtual situation display, which gives a synthetic picture of the tactical situation, or go for a visual attack using his head-up display (HUD).

The latest development of the F-15 is the F-15E Strike Eagle, which was originally developed as a private venture. The prototype first flew in 1982. The Strike Eagle carries a two-man crew, the pilot and back-seat weapons and defensive systems operator. The avionics suite is substantial, and to accommodate it one of the fuselage tanks has been reduced. More economical and reliable engines have been fitted without the need for extensive airframe modifications. Strengthened airframe and landing gear allow a greater weapons load. F-15E units were at the forefront of precision bombing during the Gulf War.

Specification: McDonnell Douglas F-15E Strike Eagle	
Type:	two-seat strike/attack aircraft and air superiority fighter
Powerplant:	two 10,779kg (23,770lb) thrust Pratt & Whitney F100-PW-220 turbofans
Performance:	maximum speed 2655km/h (1650mph) at altitude; service ceiling 30,500m (100,000ft); range 5745km (3570 miles) with conformal fuel tanks
Weights:	empty 14,375kg (31,700lb); maximum take-off 36,733kg (81,000lb)
Dimensions:	wing span 13.05m (42ft 10in); length 19.43m (63ft 9in); height 5.63m (18ft 5in); wing area 56.48m² (608 sq ft)
Armament:	one 20mm (0.79in) M61A-1 cannon; four AIM-7 or AIM-120 and four AIM-9 AAMs; many combinations of underwing ordnance

Lockheed F-117A Night Hawk

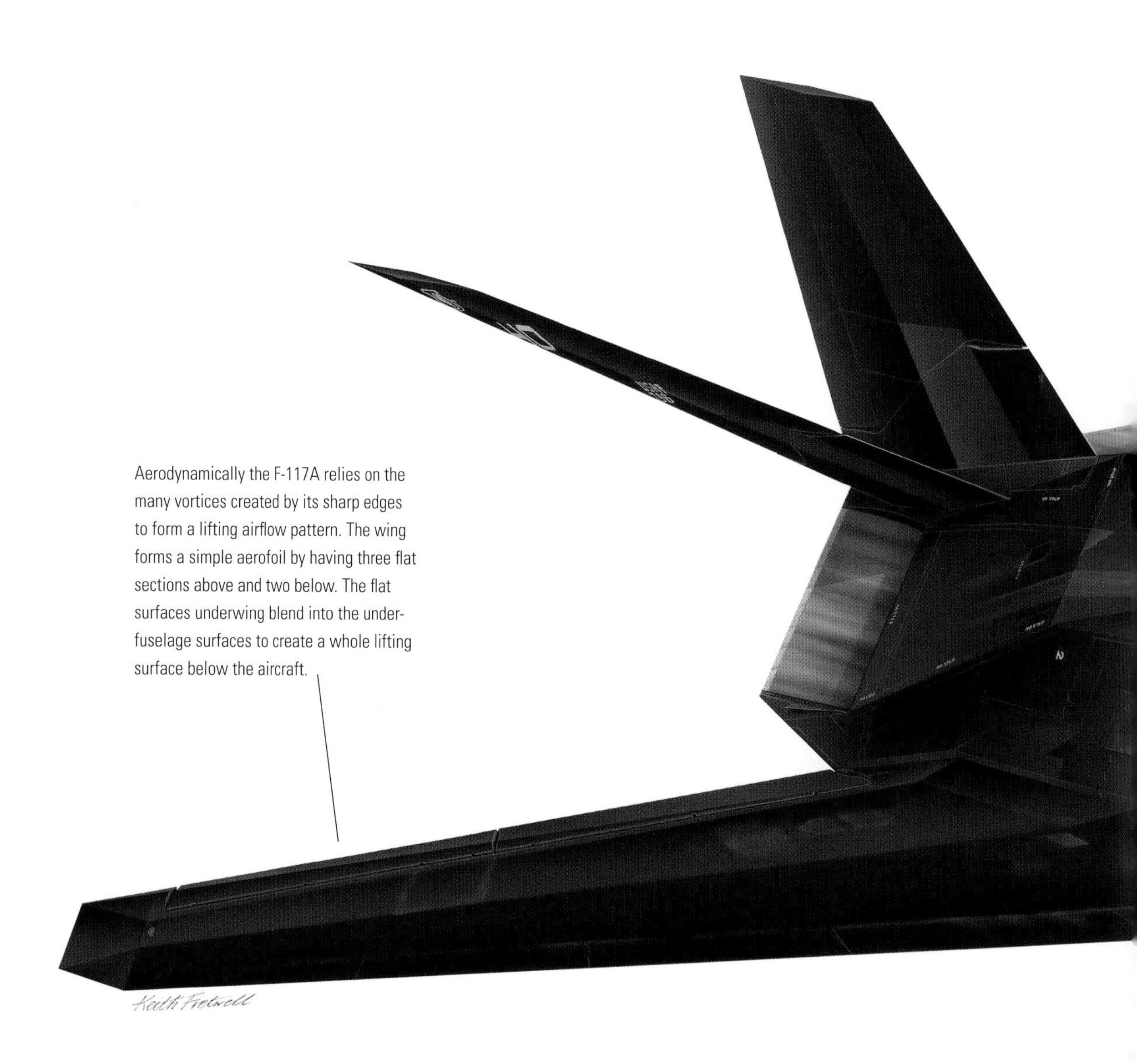

Aerodynamically the F-117A relies on the many vortices created by its sharp edges to form a lifting airflow pattern. The wing forms a simple aerofoil by having three flat sections above and two below. The flat surfaces underwing blend into the under-fuselage surfaces to create a whole lifting surface below the aircraft.

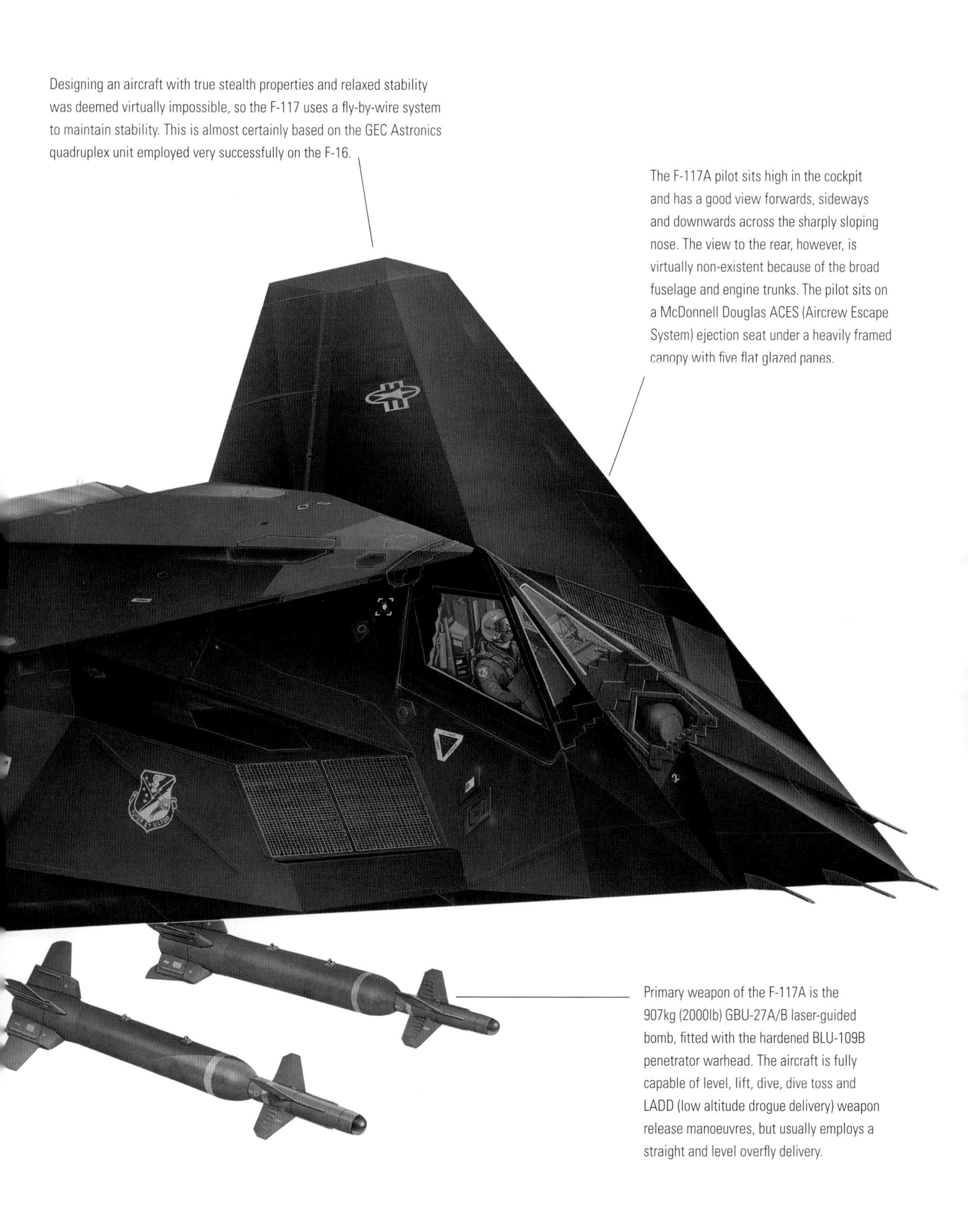

Designing an aircraft with true stealth properties and relaxed stability was deemed virtually impossible, so the F-117 uses a fly-by-wire system to maintain stability. This is almost certainly based on the GEC Astronics quadruplex unit employed very successfully on the F-16.

The F-117A pilot sits high in the cockpit and has a good view forwards, sideways and downwards across the sharply sloping nose. The view to the rear, however, is virtually non-existent because of the broad fuselage and engine trunks. The pilot sits on a McDonnell Douglas ACES (Aircrew Escape System) ejection seat under a heavily framed canopy with five flat glazed panes.

Primary weapon of the F-117A is the 907kg (2000lb) GBU-27A/B laser-guided bomb, fitted with the hardened BLU-109B penetrator warhead. The aircraft is fully capable of level, lift, dive, dive toss and LADD (low altitude drogue delivery) weapon release manoeuvres, but usually employs a straight and level overfly delivery.

The amazing F-117A 'Stealth' aircraft began life in 1973 as a project called 'Have Blue', launched to study the feasibility of producing a combat aircraft with little or no radar and infrared signature. Two Experimental Stealth Tactical (XST) 'Have Blue' research aircraft were built and flown in 1977 at Groom Lake, Nevada (Area 51). One was destroyed in an accident, but the other went on to complete the test programme successfully in 1979. The Have Blue prototypes validated the faceting (angled flat surfaces) concept of the stealth aircraft, and the basic aircraft shape. A key difference between these aircraft and the production F-117 was the inward canting of the fins, which were mounted on the outside of the main fuselage body and much further forward than on production aircraft. The leading edge was set at a very sharp 72.5 degrees. Have Blue utilized many off-the-shelf systems from other aircraft, including the fly-by-wire system of the F-16. The aircraft also had the F-16's sidestick controller, while the undercarriage came from the Northrop F-5. The two engines came from a Rockwell T-2 Buckeye. Flight control systems were served by three static pressure sensors on the forward fuselage, and three total pressure probes, one on the nose and two on the cockpit windscreen post. Have Blue 1001 also had the instrumentation boom that correlated data from the primary systems. The exhaust slot for the Have Blue had a greater extension on its lower lip than was featured on the F-117, with its two exhaust slots meeting at a common point on the centreline. The lower portion of the nozzle formed a two-position plate that automatically deflected downwards when the angle of attack exceeded 12 degrees.

The evaluation of the two Have Blue aircraft led to an order for 65 production F-117As. Five of these were used for evaluation, and one crashed before delivery. The type made its first flight in June 1981 and entered service in October 1983. The F-117A is a single-seat, subsonic aircraft powered by two non-afterburning GE F404 turbofans with shielded slot exhausts designed to dissipate heat emissions (aided also by heat-shielding tiles), thus minimizing the infrared signature. The use of faceting scatters incoming radar energy; radar-absorbent materials and transparencies treated with conductive coating reduce the F-117A's radar signature still further. The aircraft has highly swept wing leading edges, a W-shaped trailing edge, and a V-shaped tail unit. Armament is carried on swing-down trapezes in two internal bays. The F-117A has quadruple redundant

Right: **The Lockheed F-117A 'stealth' aircraft has played a vital part in the limited wars of the late 20th and early 21st centuries, from the Gulf to Kosovo. This head-on view clearly shows the heavily framed cockpit canopy and the twin trapezes which extend from the weapons bay.**

813

fly-by-wire controls, steerable turrets for FLIR and laser designator, head-up and head-down displays, laser communications and nav/attack system integrated with a digital avionics suite.

F-117As of the 37th Tactical Fighter Wing played a prominent part in the 1991 Gulf War, making first strikes on high priority targets; since then they have been used in the Balkans and Afghanistan. The last of 59 F-117As was delivered in July 1990. Along with the Northrop Grumman B-2 Spirit 'stealth bomber', the Night Hawk is the USAF's primary attack weapon, the F-117 force being able to exert an influence on an air campaign that far outweighs its

Above: **An F-117A of the 37th Tactical Fighter Wing over the Tonopah Test Range in Nevada. Its stealth features enable the F-117 to cruise into a target area undetected and unmolested by hostile air defences, and achieve the optimum position to launch an attack.**

Opposite: **This early photograph of an F-117A, issued by Lockheed, was one of the first to reveal the aircraft's extraordinary design features, the result of years of experimentation. The basic design was tested on two experimental aircraft under a project named Have Blue.**

meagre size. As was seen during Desert Storm, the F-117's primary role is to attack high-value command, control and communications targets, to, in effect, 'decapitate' the enemy's ability to control his forces. Such targets include leadership bunkers, command posts and air defence and communications centres. Most of these are well-defended, usually hardened against normal attacks and often in downtown city areas. Many may have only one or two weak spots, such as air shafts, where a bomb may do any damage at all. The need to deliver a high-energy penetration weapon causing the least possible collateral damage requires the utmost accuracy and high survivability. By using stealth, the F-117 can cruise into the target area undetected and unmolested by hostile air defences, relying on its extensive low-observable properties for protection, and achieve the optimum position for an accurate attack, whereas the crew of a conventional aircraft has to fly low and fast while dodging defences to penetrate the target area, leaving little time to concentrate on accurate weapons delivery. By contrast, the F-117 pilot can take time to use his sophisticated weapons system to ensure a pinpoint strike. If the strategic targets run out, the F-117 is a valuable weapon in a standard interdiction role, being able to hit bridges, railway depots, airfields and industrial complexes with ease.

Specification: Lockheed F-117A Night Hawk	
Type:	single-seat stealth interdictor
Powerplant:	two 4899kg (10,800lb) thrust General Electric F404-GE-F1D2 turbofans
Performance:	maximum speed Mach 0.92 at altitude; service ceiling classified; range classified
Weights:	empty approx 13,605kg (30,000lb); maximum take-off 23,810kg (52,500lb)
Dimensions:	wing span 13.20m (43ft 4in); length 20.08m (65ft 11in); height 3.78m (12ft 5in); wing area 105.9m^2 (1140 sq ft)
Armament:	provision for 2268kg (5000lb) of stores on rotary dispenser in weapons bay, including the AGM-88 HARM anti-radiation missile, AGM-65 Maverick ASM, GBU-19 and GBU-27 optronically guided bomb, BLU-109 laser-guided bombs, and B61 free-fall nuclear bomb

Rockwell B-1B Lancer

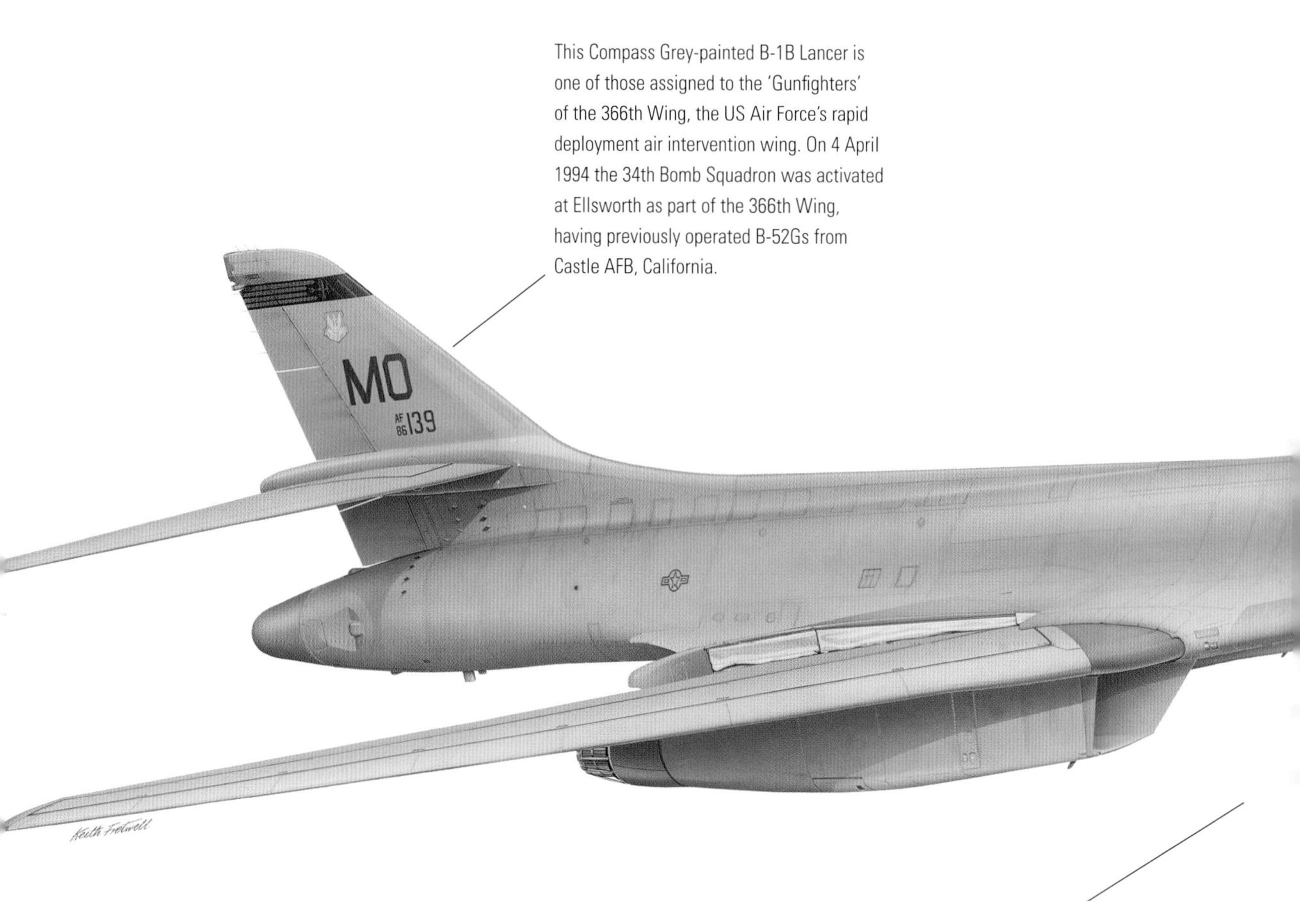

This Compass Grey-painted B-1B Lancer is one of those assigned to the 'Gunfighters' of the 366th Wing, the US Air Force's rapid deployment air intervention wing. On 4 April 1994 the 34th Bomb Squadron was activated at Ellsworth as part of the 366th Wing, having previously operated B-52Gs from Castle AFB, California.

The B-1B can carry a Conventional Weapon Module (CWM) in each of its three weapons bays. The CWM does not rotate, but is a rigid frame carried on the same trunnions as the rotary launchers for the air-launched cruise missiles. Development of the CWM made it easier to load conventional bombs.

The B-1B's flight refuelling receptacle is situated directly in front of the cockpit, making the establishment of a good contact much easier for the pilots. The white markings give the boom operator in the tanker aircraft a clear guide when refuelling at night.

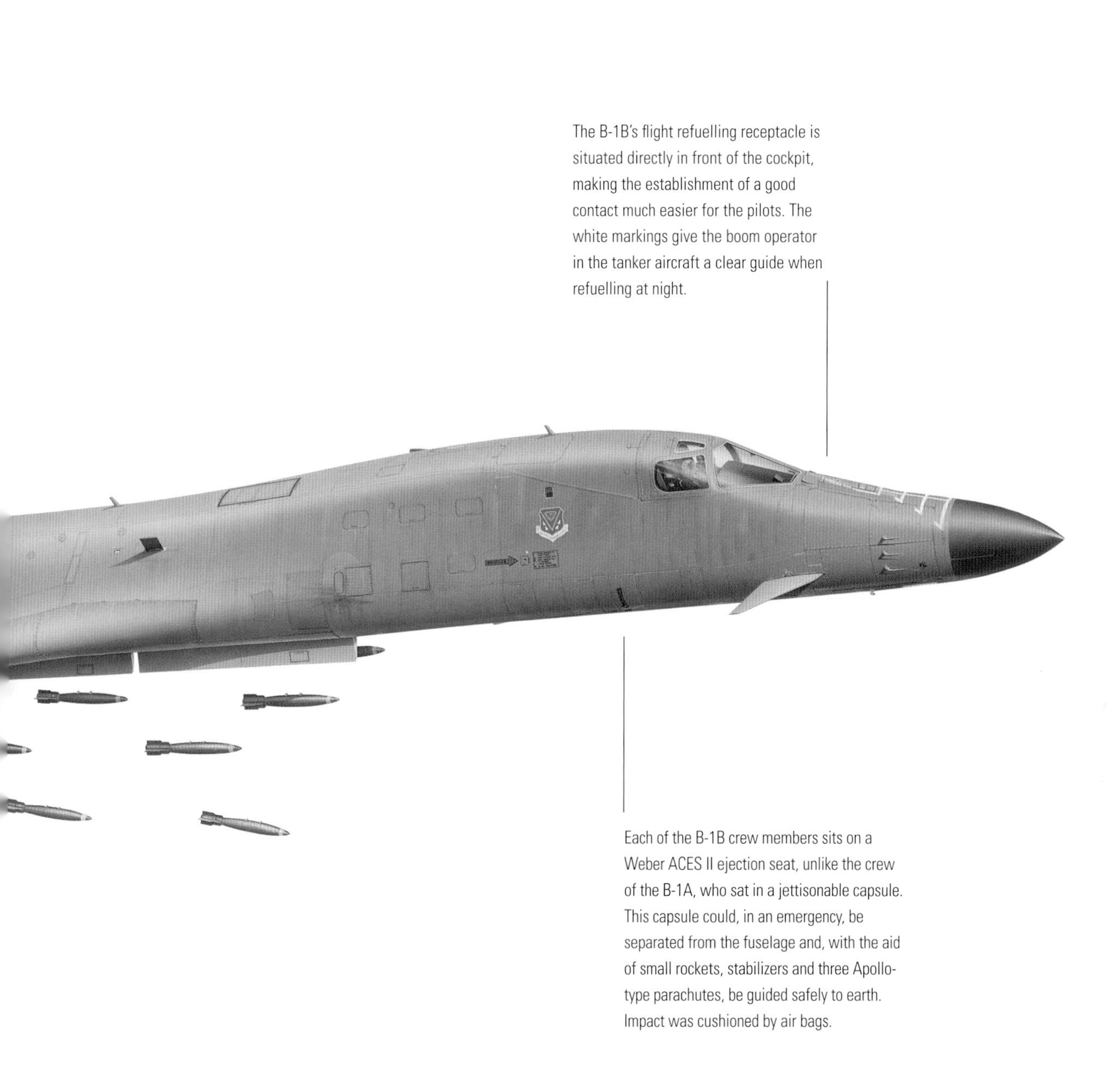

Each of the B-1B crew members sits on a Weber ACES II ejection seat, unlike the crew of the B-1A, who sat in a jettisonable capsule. This capsule could, in an emergency, be separated from the fuselage and, with the aid of small rockets, stabilizers and three Apollo-type parachutes, be guided safely to earth. Impact was cushioned by air bags.

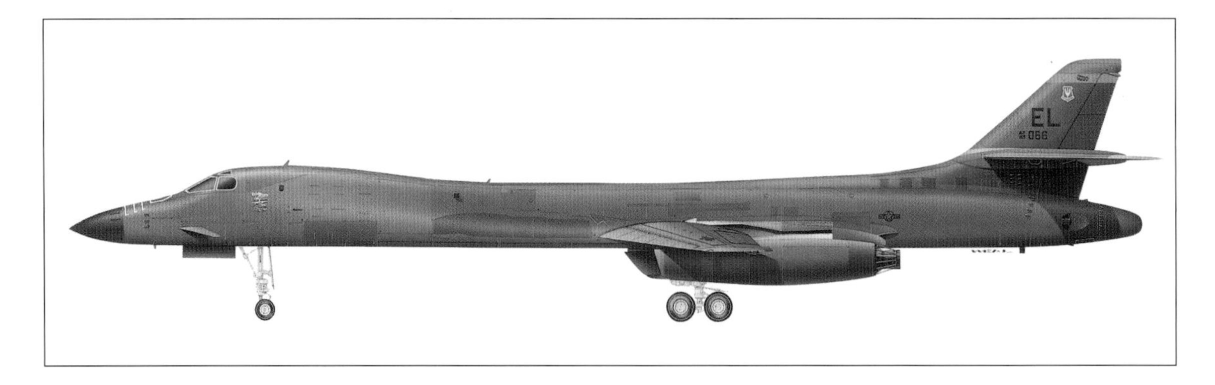

Above: **The Rockwell B-1B flew for the first time in October 1984 and was well ahead of schedule, despite the crash several weeks earlier of one of the two B-1A prototypes allocated to the test programme. The first operational B-1B was delivered on 7 July 1985.**

Designed to replace the B-52 and FB-111 in the low-level penetration role, the B-1 prototype flew on 23 December 1974, and subsequent flight trials and evaluation progressed rapidly. On 21 April 1975 a Strategic Air Command KC-135 tanker crew of the 22nd ARS conducted the first flight refuelling trials with the new bomber, and on 19 September it was flown for the first time by an SAC pilot, Major George W. Larson of the 4200th Test and Evaluation Squadron at Edwards AFB. Major Larson handled the controls for about one-third of the six-and-a-half-hour flight.

Trials continued throughout the following year, and on 2 December 1976 the US Secretary of Defense, Donald H. Rumsfeld, after consultations with President Gerald Ford, authorized the USAF to proceed with production of the B-1. In September, however, Congress had restricted funding of the B-1 programme to $87 million per month, slowing down the programme and effectively leaving the decision on the B-1's future to President Jimmy Carter, who would take office on 20 January 1977.

The decision hung in the balance until 30 June 1977, when President Carter delivered his bombshell and stated in a nationwide TV address that the B-1 would not be produced. On 2 October 1981, however, President Ronald Reagan's new US Administration took the decision to resurrect the Rockwell B-1 programme. Between 1977 and 1981 the USAF had used the B-1 prototype in a bomber penetration evaluation, and this had resulted in a unique opportunity to rate the combat effectiveness of an advanced bomber already cancelled as a production programme, with no pressure to prove the case one way or the other. The conclusion reached was that, with skilled crews and flexible tactics, the bombers were getting through to their targets more often than the computers had predicted, a fact that was firmly presented in a report to Congress submitted early in 1981.

The operational designation of the supersonic bomber, 100 of which were to be built for SAC, was to be B-1B, the prototypes already built now being known as B-1As. The primary mission of the B-1B would be penetration with free-fall weapons, using SRAMS for defence suppression. The aircraft would also be modified to carry the ALCM, being fitted with a movable bulkhead between the two forward bomb bays to make room for an ALCM launcher.

The first B-1B flew in October 1984 and was well ahead of schedule, despite the crash several weeks earlier of one of the two B-1A prototypes taking part in the test programme. The first operational B-1B (83-0065) was delivered to the

Specification: Rockwell B-1B

Type:	four-crew strategic bomber
Powerplant:	four 13,958kg (30,780lb) thrust General Electric F101-GE-102 turbofans
Performance:	maximum speed 1328km/h (825mph) at high altitude; service ceiling 15,240m (50,000ft); range 12,000km (7455 miles) on internal fuel
Weights:	empty 87,072kg (192,000lb); maximum take-off 216,319kg (477,000lb)
Dimensions:	wing span 41.65m (136ft 8in); length 44.81m (147ft); height 10.36m (34ft); wing area 181.16m² (1950 sq ft)
Armament:	up to 38,320kg (84,500lb) of Mk 82 or 10,974kg (24,200lb) of Mk 84 iron bombs in the conventional role, 24 SRAMs, twelve B.28 and B.43 or 24 B.61 and B.83 free-fall nuclear bombs, eight ALCMs on internal rotary launchers and 14 more on underwing launchers, and various combinations of other underwing stores. Low-level operations are flown with internal stores only

96th Bomb Wing at Dyess AFB on 7 July 1985 – although in fact it was the fleet prototype, 82-0001, that underwent the SAC acceptance ceremony, the other aircraft having suffered engine damage from ingesting nuts and bolts from a faulty air conditioner.

Despite a series of problems with avionics and systems, B-1B deliveries to SAC reached a rate of four per month in 1986. In January 1987 the trials aircraft successfully launched a SRAM for the first time, and in April an aircraft from the 96th BW completed a mission lasting 21 hours and 40 minutes, which involved five in-flight refuellings to maintain a high all-up weight, the aircraft flying at about 741km/h (460mph) and covering 15,138km (9407 miles). This operation was in connection with the development of operational techniques involving the carriage of very heavy loads over long distances. Most B-1B missions are flown at high subsonic speeds; the aircraft is fitted with fixed-geometry engine inlets that feed the engines through curved ducts incorporating stream-wise baffles, blocking radar reflections from the fan. These reduce the maximum speed to Mach 1.2; the earlier B-1A had external-compression inlets and could reach Mach 2.2, but it had about ten times the B-1B's radar signature.

A good deal of so-called 'stealth' technology has been built into the B-1B, greatly enhancing its prospects of penetrating the most advanced enemy defences.

Below: **A Rockwell B-1B of the 28th Bomb Wing. The primary mission of the B-1B was defined as the penetration of enemy territory with free-fall weapons, using short-range attack missiles (SRAM) for defence suppression. The B-1B can also carry ALCM cruise missiles.**

Northrop Grumman B-2 Spirit

In addition to the shape of the aircraft, a key component of the B-2's radar signature reduction system is the use of special coatings that absorb radar energy and transmit it round the aircraft's surface. The application of such materials requires attention to minute details, and the materials themselves suffer from adverse climatic conditions.

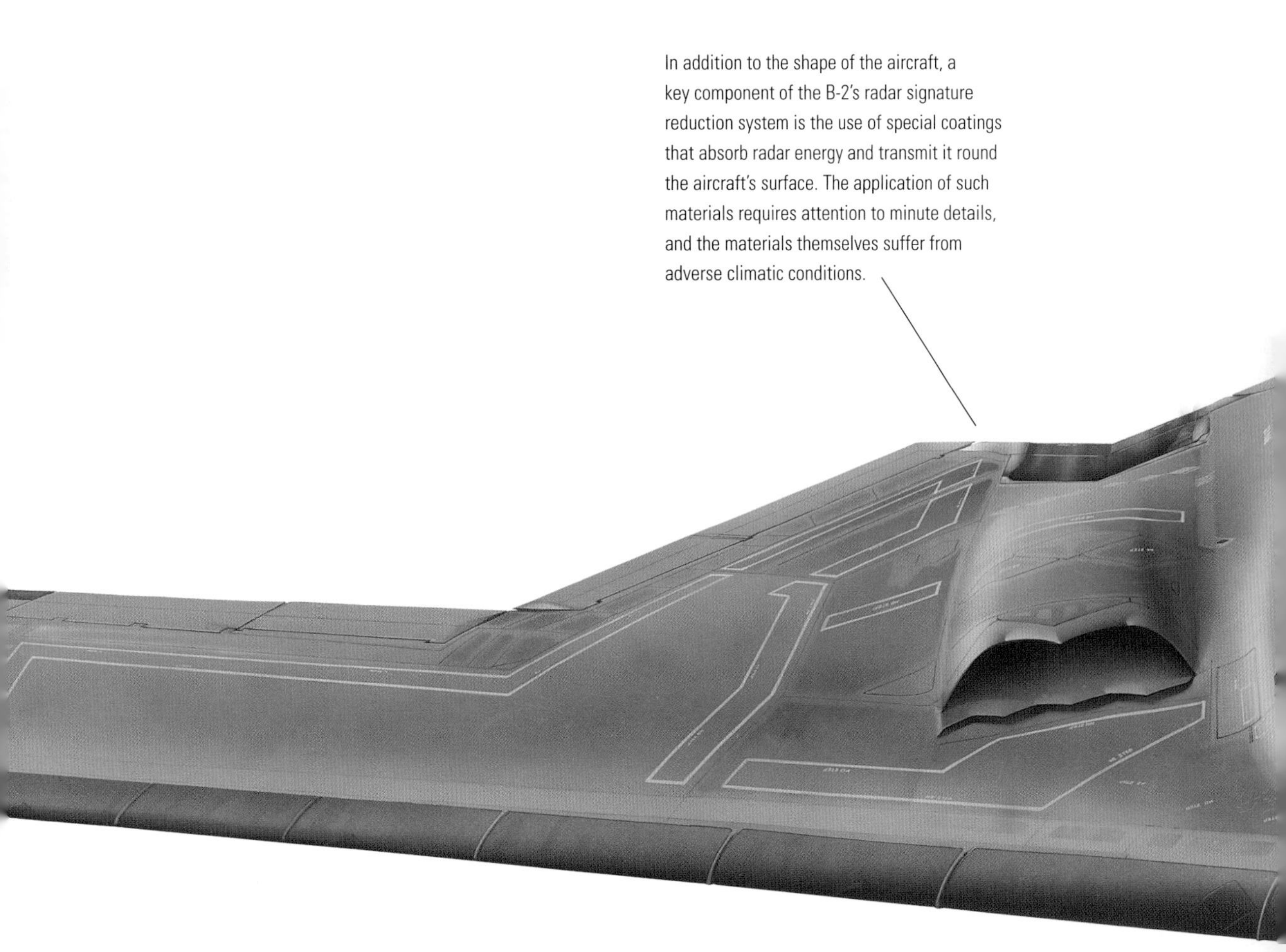

The B-2's designers realized that the problem of combining stealth with doors and other apertures could be eased if the door edges were serrated, so that the edges would be angled while the apertures themselves were rectangular. In 1977 Northrop created a shaping technique that combined sharp edges with curved surfaces.

The flying wing was chosen for the B-2 because of its stealth properties, its basic mission and the intended mission profile. The flat, low profile shape with no tail surfaces enhances the stealth characteristics. As a long-range bomber, the B-2 can carry a dense, compact payload that would fit inside the wing.

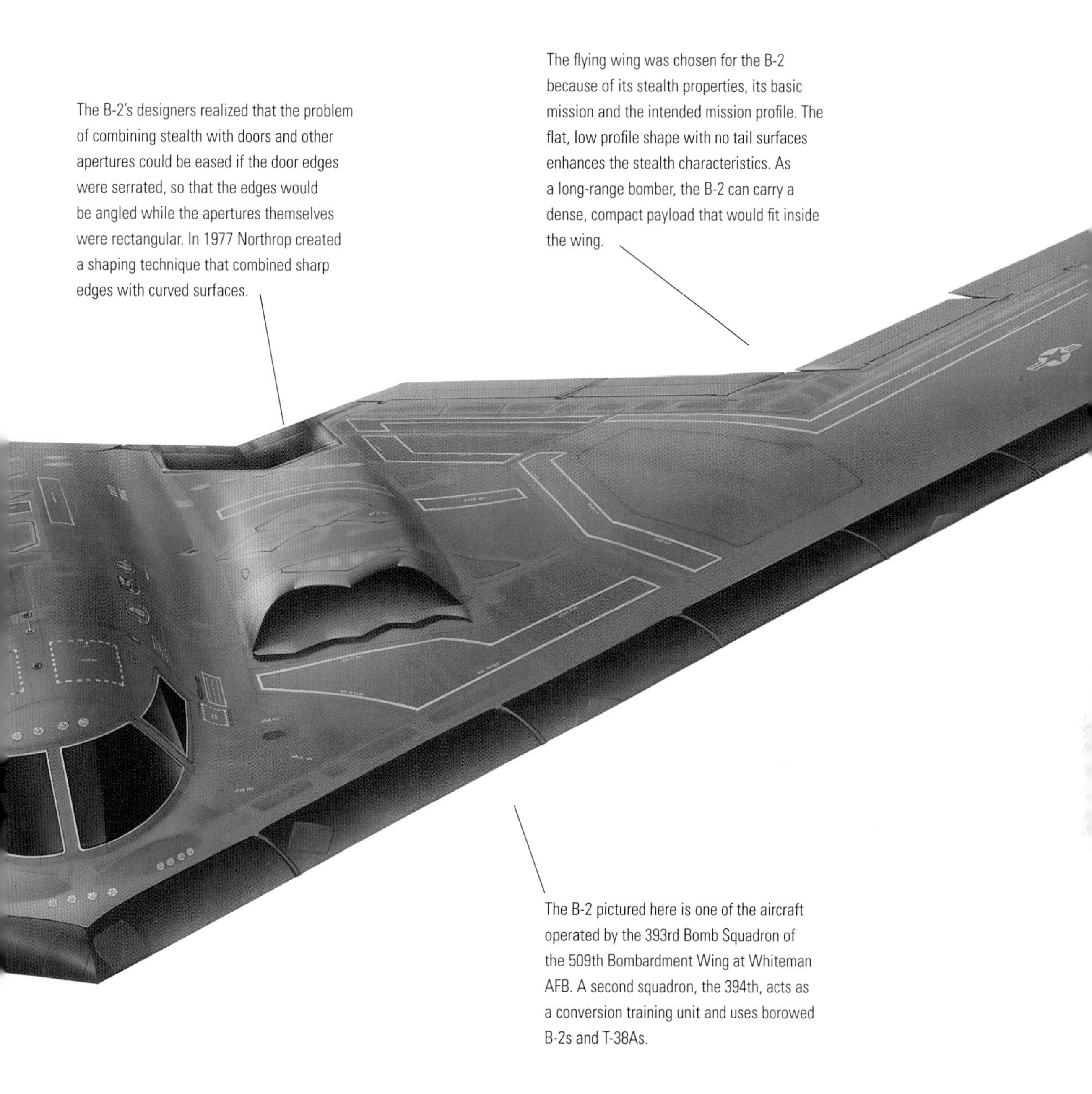

The B-2 pictured here is one of the aircraft operated by the 393rd Bomb Squadron of the 509th Bombardment Wing at Whiteman AFB. A second squadron, the 394th, acts as a conversion training unit and uses borowed B-2s and T-38As.

In modern air warfare the dividing line between strategic and tactical air power has become somewhat blurred, with strategic aircraft sometimes being used in a tactical role. The saturation bombing of Iraqi troops by B-52s in the Gulf War is just one example. But one aircraft type in particular will have a clearly defined strategic role in the twenty-first century. This is the Northrop B-2 Spirit strategic penetration bomber, the embodiment of 'stealth' technology pioneered operationally by the Lockheed F-117A fighter-bomber. Development of the B-2 was begun in 1978 and the USAF originally wanted 133 examples, but by 1991 successive budget cuts had reduced this to 21 aircraft. The first B-2 (880329) was delivered to the 393rd Bomb Squadron of the 509th Bomb Wing at Whiteman AFB, Missouri, on 17 December 1993, with a second squadron, the 715th BS, also scheduled to equip, bringing the 509th BW's establishment up to 16 aircraft. Ever since the 509th BW was formed in 1944 to train for dropping the two atomic bombs on Japan, it has had a specialist role in introducing new types of bomber aircraft into operational

Opposite: **A B-2 Spirit stealth bomber of the 509th Bomb Group pictured on a training sortie from Whiteman AFB. Ever since the 509th BG was formed in 1944 to drop the atomic bombs on Japan, it has pioneered the USAFs operational bombing tactics and deployment of new weapons.**

service and pioneering new tactics. The B-2, which is powered by four 7847kg (17,300lb) thrust General Electric F118-GE-100 non-afterburning turbofans, has two weapons bays mounted side by side in the lower centrebody, each fitted with a Boeing rotary launcher assembly. The bomb cells can accommodate 16 AGM-129 Advanced Cruise Missiles, or alternatively 16 B.61 or B.83 free-fall nuclear bombs, 80 Mk 82 227kg (500lb) bombs, 16 Joint Direct Attack Munitions, 16 Mk 84 906kg (2000lb) bombs, 36 M117 340kg (750lb) fire bombs, 36 CBU-87/89/97/98 cluster bombs, and 80 Mk 36 254kg (560lb) or Mk 62 sea mines. With a typical weapons load, the B-2 has a range of 12,045km (6500nm) at high level and 8153km (4400nm) at low level.

The primary weapon in the B-2's nuclear arsenal is the variable-yield B.83 megaton-range bomb. The weapon is cleared for carriage by a variety of aircraft, but its high yield makes it more applicable for carriage by the B-2 and previously by the Rockwell B-1B. Developed as a cheaper alternative to the B.77, the B.83 emerged with similar characteristics, and was the first strategic yield weapon designed for low-level laydown deliveries, replacing the B.28, B.43 and B.57 weapons. It can be deployed from as low as 46m (150ft) and has fully variable fuzing and yield, this being programmed by the crew in flight. The B.83 is primarily targeted against hardened military objectives such as ICBM silos, underground facilities and nuclear weapons storage facilities.

In designing the Advanced Technology Bomber (ATB), as the B-2 project was originally known, the Northrop Company decided on an all-wing configuration from the outset. Flying wing devotees such as Hugo Junkers and Jack Northrop have existed as long as aviation itself, arguing that a flying wing will carry the same payload as a conventional aircraft while weighing less and using less fuel. The weight and drag of the tail surfaces are absent, as is the weight of the structure that supports them. The wing structure itself is far more efficient because the weight of the aircraft is spread across the wing rather than concentrated in the centre. Northrop's experimental piston-engined flying wing bomber of the 1940s was designed to equal the range and carry the same warload as the Convair B-36, but at two-thirds the gross weight and two-thirds the power. The company also produced a prototype flying wing jet bomber, the YB-49, in 1947, but this had little influence on the decision to pursue an all-wing solution for the B-2; the all-wing approach was selected because it promised to result in an exceptionally clean configuration for minimizing radar cross-section, including the elimination of vertical tail surfaces, with added benefits such as span-loading structural efficiency and high lift/drag ratio for efficient cruise. Outboard wing panels were added for longitudinal balance, to increase lift/drag ratio and to provide sufficient span for pitch, roll and yaw control. Leading-edge sweep was selected for balance and transonic aerodynamics, while the overall planform was designed to have neutral longitudinal (pitch) static stability. Because of its short length, the aircraft had to produce stabilizing pitchdown moments beyond the stall for positive recovery. The original ATB design had elevons on the outboard wing panels only, but as the design progressed additional elevons were added inboard, giving the B-2 its distinctive 'double-W' trailing edge. The wing leading edge is so designed that air is channelled into the engine intakes from all directions, allowing the engines to operate at high power and zero airspeed. In transonic cruise, air is slowed from supersonic speed before it enters the hidden compressor faces of the GE F118 engines.

A stores management processor handles the B-2's 22,730kg (50,120lb) weapons load. A separate processor controls the Hughes APQ-181 synthetic-aperture radar and its input to the display processor. The radar has 21 operational modes, including high-resolution ground mapping.

The aircraft is highly manoeuvrable, with fighter-like handling characteristics.

Specification: Northrop B-2 Spirit

Type:	four-crew strategic bomber
Powerplant:	four 7847kg (17,300lb) thrust General Electric F118-GE-110 turbofans
Performance:	maximum speed 764km/h (475mph) at high altitude; service ceiling 15,240m (50,000ft); range 11,675km (7255 miles)
Weights:	empty 45,350kg (100,000lb); maximum take-off 181,400kg (400,000lb)
Dimensions:	wing span 52.43m (172ft); length 21.03m (69ft); height: 5.18m (17ft); wing area approx 463.50m² (5000 sq ft)
Armament:	sixteen AGM-129 Advanced Cruise Missiles, or alternatively 16 B.61 or B.83 free-fall nuclear bombs, 80 Mk 82 227kg (500lb) bombs, 16 Joint Direct Attack Munitions, sixteen Mk 84 906kg (2000lb) bombs, 36 M117 340kg (750lb) fire bombs, 36 CBU-87/89/97/98 cluster bombs, and 80 Mk 36 254kg (560lb) or Mk 62 sea mines

Lockheed Martin F-22 Raptor

A Raptor pilot can receive information from other F-22s, allowing a radar-silent attack. A Raptor that is outside its missile envelope can thus track a target and covertly send target data to a closer Raptor to make the silent kill.

Exploiting its stealthy characteristics, the F-22 has a 'first look, first shot' advantage. The AN/APG-77 active electronically scanned array (AESA) radar can track targets before going electronically silent.

The two Pratt & Whitney F119 engines allow the Raptor to accelerate to and cruise at speeds of about Mach 1.8 without using afterburners.

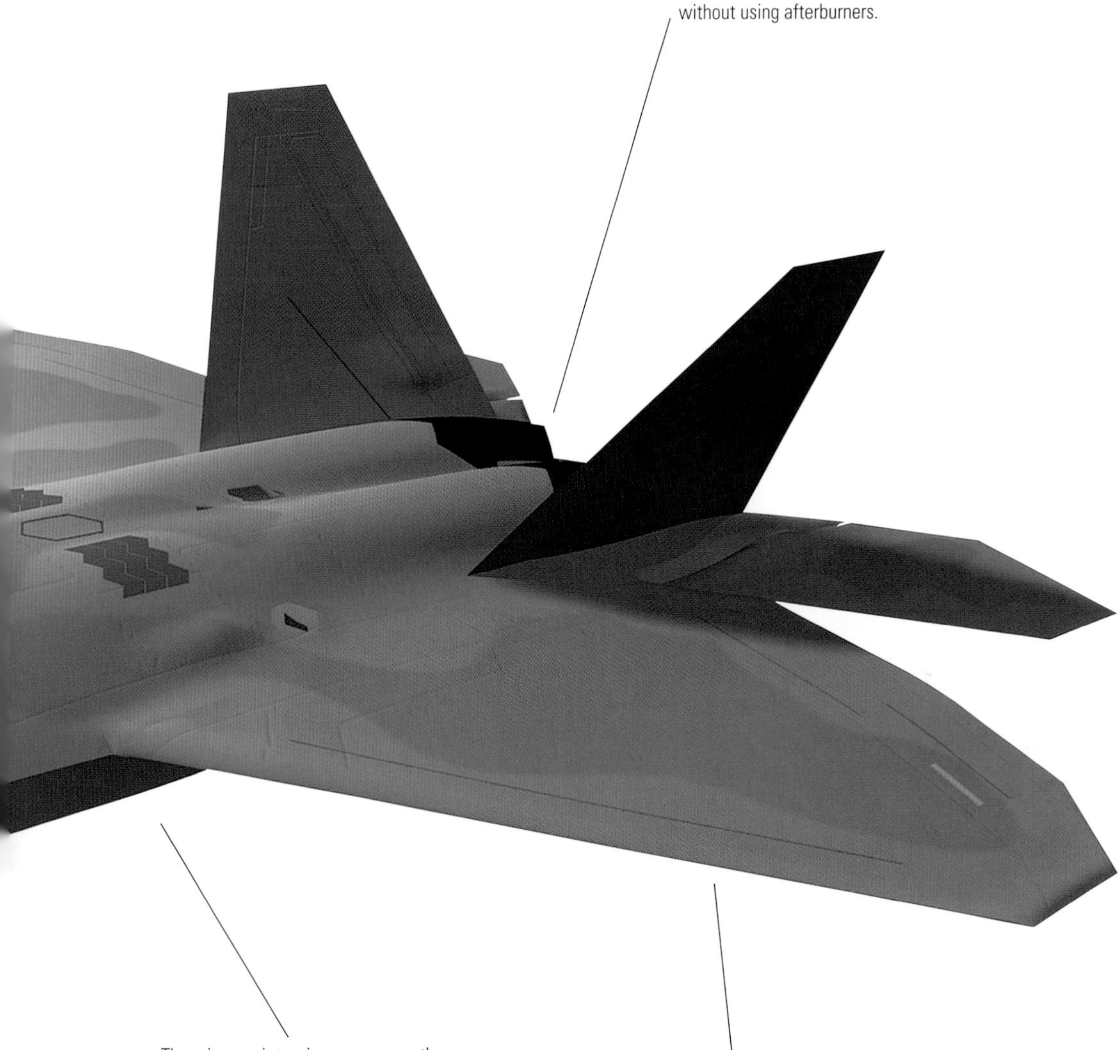

The primary air-to-air weapons are the AIM-9M Sidewinder and the AIM-120C AMRAAM. Weapons are carried in the internal weapons bays, which open up at the very last second as the Raptor 'uncloaks'.

Advanced aerodynamics, combined with thrust vectoring and cutting-edge flight control systems, provide so-called 'super manoeuvrability'.

Above: **An F-22A of 27th Fighter Squadron, 1st Fighter Wing, the USAF's first Raptor unit.**

Widely regarded as the most capable air superiority fighter in service anywhere in the world, the F-22A is capable of both air-to-air and air-to-ground missions, and has been designed to combine stealth, performance, agility and integrated avionics in a single airframe.

The Raptor is described by its operator as representing 'an exponential leap in warfighting capabilities'. The F-22A began life as the Advanced Tactical Fighter (ATF) programme which entered the demonstration and validation phase in 1986. The U.S. Air Force lined up seven companies as contenders for the ATF programme, with Lockheed and Northrop as the primary airframe contractors. Lockheed teamed up with Boeing and General Dynamics, while Northrop's effort was joined by principle contractor McDonnell Douglas. The resulting YF-22 and YF-23 prototypes were built for a fly-off competition, which also involved rival engines from Pratt & Whitney and General Electric (the YF119 and YF120, respectively). The YF-22 made its first flight in September 1990.

The USAF selected the YF-22 and YF119 as the most promising and launched the engineering and manufacturing development (EMD) phase in 1991. Development contracts were issued to Lockheed/Boeing (airframe) and Pratt & Whitney (engines). EMD included extensive tests of systems and subsystems, as well as flight testing with nine aircraft at Edwards Air Force Base, California. A first EMD flight was recorded in 1997.

In 2001 the programme received approval to enter low-rate initial production. There followed initial operational and test evaluation by the Air Force Operational Test and Evaluation Centre, which was successfully concluded in 2004. Approval for full-rate production was granted in 2005. After briefly receiving the F/A-22 designation in recognition of its attack capabilities, the Raptor was renamed as the F-22A in December 2005; at the same time the aircraft achieved initial operational capability with the USAF's 27th Fighter Squadron. At one time, the USAF planned to procure a minimum of 750 ATFs, but the end of the Cold War saw this dramatically scaled back, and finally just 187 aircraft were acquired at a unit cost of $143 million.

The Raptor's prowess in the air-to-air arena is ensured through a combination of sensor capability, integrated avionics, situational awareness and weaponry. Combined with its stealth characteristics and performance, the sensor suite is intended to permit the F-22 pilot to track, identify, engage and kill air-to-air threats before being detected. Primary sensors are the AN/APG-77 active electronically scanned array (AESA) radar and AN/ALR-94 passive

Above: **F-22 Raptor releases a 450kg (1000lb) bomb over the Utah Test and Training Range during weapons evaluation.**

receiver system. A high degree of situational awareness is ensured through advanced cockpit design and sensor fusion. In typical air-to-air configuration the Raptor is armed with six AIM-120C AMRAAMs and two AIM-9M Sidewinder missiles.

To ensure its own protection while in flight, the F-22A also relies on low-observable 'stealth' technologies. The aircraft can also call upon its sparkling performance to evade air-to-air and surface-to-air threats. Power is provided by a pair of Pratt & Whitney F119-PW-100 turbofan engines, which produce more thrust than any other current fighter engine. The F-22A is able to cruise at supersonic airspeeds without using afterburner – a characteristic known as supercruise. Supercruise also offers the advantage in conserving fuel, providing an increase in endurance and range. The engines are equipped with thrust vectoring, combined with advanced aerodynamics and flight controls, and a high thrust-to-weight ratio to provide excellent agility.

Although it was originally planned as an air dominance fighter, the F-22 has latterly emerged as a powerful attack aircraft. In the air-to-ground configuration the aircraft can carry two 907kg (1000lb) GBU-32 Joint Direct Attack Munitions (JDAM) internally. Work is underway to improve the air-to-ground potential of the Raptor, adding radar modifications and the capability to carry eight Small Diameter Bombs. Whether carrying JDAMs or SDBs, the Raptor can also carry two AIM-120s and two AIM-9s for self defence. In this role the F-22A made its combat debut, striking targets in Syria in September 2014.

Specification: Lockheed Martin F-22 Raptor	
Type:	multi-role air dominance fighter
Powerplant:	two 155.69kN (35,000lb) Pratt & Whitney F119-PW-100 afterburning turbofans
Performance:	maximum speed 2410km/h (1500mph) at altitude; range: more than 2977km (1850 miles), ferry with 2 x external wing fuel tanks
Weight:	38,000kg (83,500lb) maximum take-off
Dimensions:	wing span 13.6m (44ft 6in); length 18.9m (62ft 1in); height: 5.1m (16ft 8in)
Armament:	one 20mm (0.79in) rotary cannon, plus two AIM-9 AAMs and six AIM-120 AMRAAMs; or (ground attack) two GBU-32 JDAMs and two AIM-120s

Lockheed Martin F-35 Lightning II

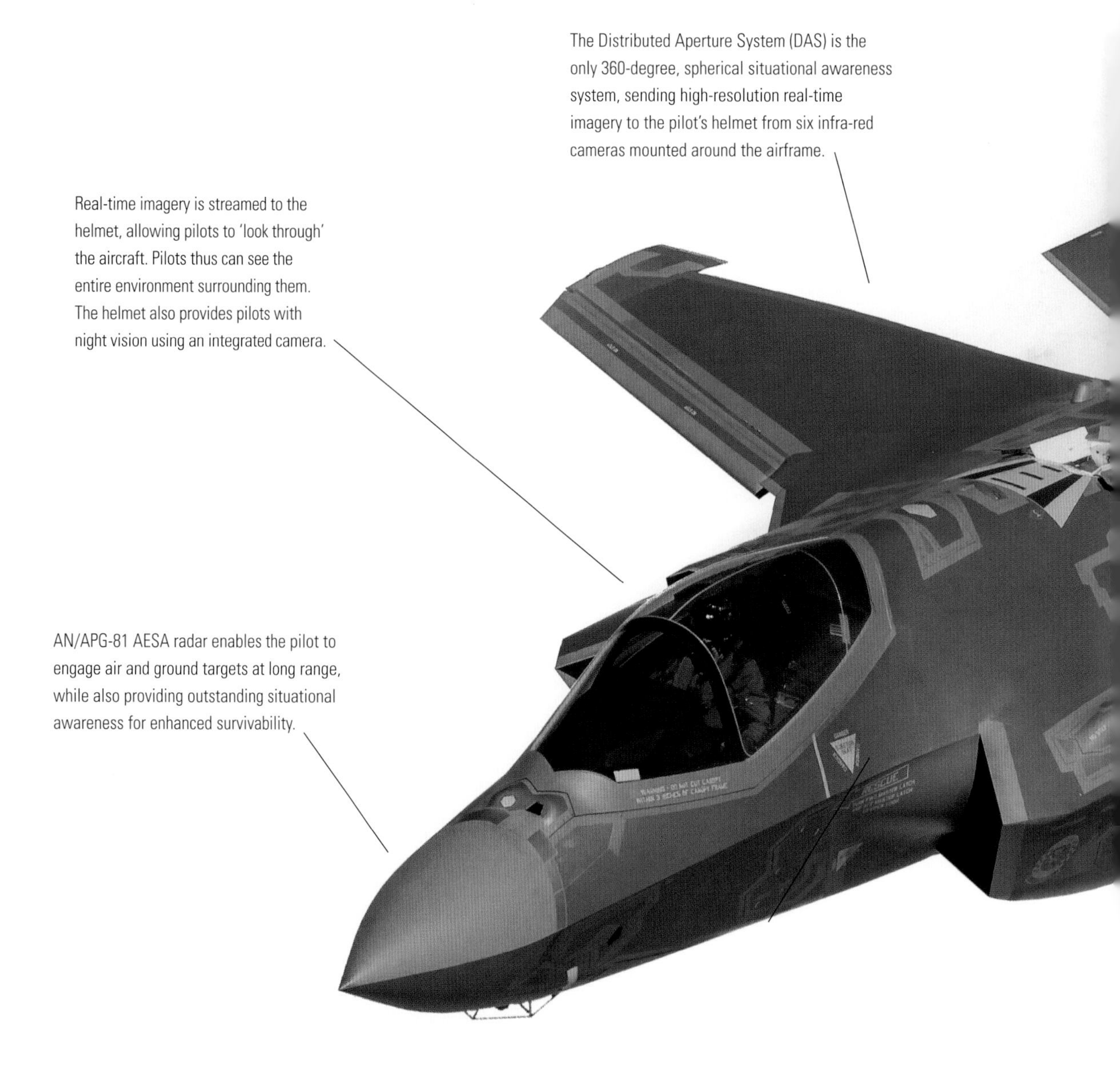

Two primary components provide vertical lift for hover: the LiftFan and 3-Bearing Swivel Module (3BSM). The LiftFan is mounted behind the cockpit. As the aircraft transitions to hover, doors open on top of the aircraft and two counter-rotating fans blow unheated air straight down, producing around half the downward thrust needed. The majority of the remaining vertical thrust is provided by the 3BSM at the rear of the aircraft.

The internal weapons bay can be complemented by external loads, for example adding four additional JDAMs and two AIM-9X AAMs underwing, complementing two internal JDAMs and two AIM-120s.

107

The F-35 is a remarkable and ambitious combat programme that combines stealth, sensor fusion and network-enabled operations in a single airframe. It is built in three variants to replace four front-line types with the US Air Force, Navy and Marine Corps, plus a variety of fighters for at least 10 other countries.

Alongside Boeing, Lockheed Martin was selected to participate in the Joint Strike Fighter (JSF) concept demonstration phase in 1997. After prototypes of the Boeing X-32 and Lockheed Martin X-35 had undergone evaluation, the Lockheed Martin design emerged victorious in October 2001. The winning company then joined forces with Northrop Grumman and BAE Systems to begin the production phase.

A first production example of the conventional take-off and landing (CTOL) F-35A was completed at Fort Worth, Texas, in February 2006. Later that year, the F-35 Joint Strike Fighter was named Lightning II. Alongside the F-35A for the US Air Force and most of the foreign operators, Lockheed Martin is building the F-35B short take-off/vertical landing (STOVL) variant for the US Marine Corps, the UK and Italy, and the F-35C carrier variant (CV) for the US Navy.

The F-35A is the only version to carry an internal cannon – the 25mm (0.98in) GAU-22/A – and will be the most prolific model, replacing the A-10 and F-16 with the USAF and serving with the majority of allied air forces to fly the Lightning II. The F-35B, which is replacing the AV-8B and F/A-18 with the USMC, is capable of STOVL operation thanks to its shaft-driven Rolls-Royce LiftFan propulsion system and an engine that can swivel 90 degrees when in short take-off/vertical landing mode. The F-35B has a smaller internal weapon bay and reduced internal fuel capacity compared to the F-35A. It is also equipped for the probe-and-drogue method of aerial refuelling.

The F-35C will replace the 'legacy' F/A-18 with the US Navy and features larger wings and strengthened undercarriage in order to cope with catapult launches and arrested landings. The CV model has folding wingtips and the greatest internal fuel capacity of the three F-35 variants, carrying nearly 9072kg (20,000lb) of internal fuel for longer range. The F-35C also uses probe-and-drogue refuelling.

At the heart of the Lightning II's capabilities are the various mission systems that include electronic sensors, displays and communications systems that collect and share data with the pilot and other friendly aircraft. Individual components include the AN/APG-81 active electronically scanned array (AESA) radar; the AN/AAQ-37 Distributed Aperture System (DAS), providing 360-degree situational awareness; the Electro-Optical Targeting System (EOTS) that combines forward-looking infra-red (FLIR) and

Below: **An F-35B Lightning II assigned to Marine Fighter Attack Squadron (VMFA) 121 prepares to make a vertical landing at Marine Corps Air Station Yuma, Ariz.**

Above: **In this photo over Fort Worth, Texas, a USAF F-35 Lightning II test aircraft AA-1 undergoes a flight check.**

infra-red search and track (IRST) functions; the Helmet-Mounted Display System; and the Communications, Navigation and Identification (CNI) system. Data from all these sensors are brought together under the sensor fusion concept, providing the pilot with a single integrated picture of the battlefield. Information can also be shared with other pilots and other assets using datalinks, such as the Multifunction Advanced Data Link (MADL).

Low observable, or stealth, features of the F-35 include the integrated airframe design, use of advanced materials, sophisticated countermeasures and provision of an on-board electronic attack capabilities. As well as the F-35's external shape, weapons can be carried internally and mission systems sensors are carefully embedded around the airframe.

The AN/APG-81 is designed to be capable of stand-off jamming for other aircraft – with a claimed 10-times increase in effective radiated power over previous fighters – and the F-35 is expected to use its stealth and survivability to operate in closer proximity to a threat from where it can provide powerful 'stand-in' jamming.

In December 2006, the F-35 completed its first flight. In the years that followed, Lockheed Martin completed flight and ground test articles of all three variants. The first production F-35 completed its maiden flight in February 2011 and deliveries began the same year.

The first of a planned 1763 F-35As for the USAF was accepted at Edwards Air Force Base, California, in May 2011. In January 2012 the first two examples of the F-35B variant was delivered to the US Marine Corps' Marine Fighter/Attack Training Squadron (VMFAT) 501 at Eglin Air Force Base, Florida. In total, the USMC plans to acquire 340 F-35Bs and 80 F-35Cs.

A first international Lightning II delivery occurred in July 2012, with the initial recipient being the United Kingdom. The F-35B was flown to Eglin AFB for operational test and evaluation, as well as training. The first of a planned 480 F-35Cs was delivered to the US Navy's Strike Fighter Squadron (VFA) 101 at Eglin in June 2013.

Specification: Lockheed Martin F-35 Lightning II	
Type:	CTOL air dominance and strike fighter
Powerplant:	one 178kN (40,000lb) Pratt & Whitney F135-PW-100 afterburning turbofan
Performance:	maximum speed 1931km/h (1200mph) with full internal weapons load; range: 2200km (1367 miles), internal fuel
Weight:	31,750kg (70,000lb) maximum take-off
Dimensions:	wing span 10.7m (35ft 0in); length: 15.7m (51ft 5in); height: 4.38m (14ft 5in)
Armament:	one 25mm (0.98in) rotary cannon, two AIM-120C AAMs, two 907kg (2000lb) GBU-31 JDAMs

United Kingdom

The United Kingdom, battered by a succession of economic crises in the years after the war, nevertheless remained at the forefront of advanced aerodynamic research and engine technology until 1950, when work slowed down dramatically.

Economic restraints imposed a heavy burden on the industry, and much of the available funding was devoted to the creation of an independent nuclear deterrent. Even so, British designers were still capable of producing aircraft that were more than a match for those of other nations, and there would undoubtedly have been a dramatic upturn in the industry's fortunes if the Government had made the right decisions in 1956–7 and not adopted the disastrous policy that crystallized in the TSR-2. The main problem that beset Britain in the early days of the jet age, however, was that there was little attempt to centralize effort or to pool knowledge. Admittedly, designers made use of the growing research facilities available, but their work progressed along independent lines, often with a wasteful duplication of effort. TSR-2 was a case in point; the fact that it reached an advanced stage of development, with a prototype flying, was little short of miraculous. But its cancellation produced one bonus: it pushed the British aerospace industry down the road of international collaboration.

Left: **The last Vulcan built, B.2 XM657, casts its shadow alongside that of the Jet Provost photographic aircraft. XM657 was delivered to the RAF in December 1964 and ended its days at RAF Manston, Kent, in 1982, where it was used for fire-fighting practice.**

De Havilland Vampire

Detachable just aft of the wing trailing edge, the tail booms were of simple semi-monocoque construction. Control cable runs were contained within each boom, and bumpers were mounted under the fins to protect the tail in the event of over-rotation. The two-spar tailplane mounted a full-span elevator.

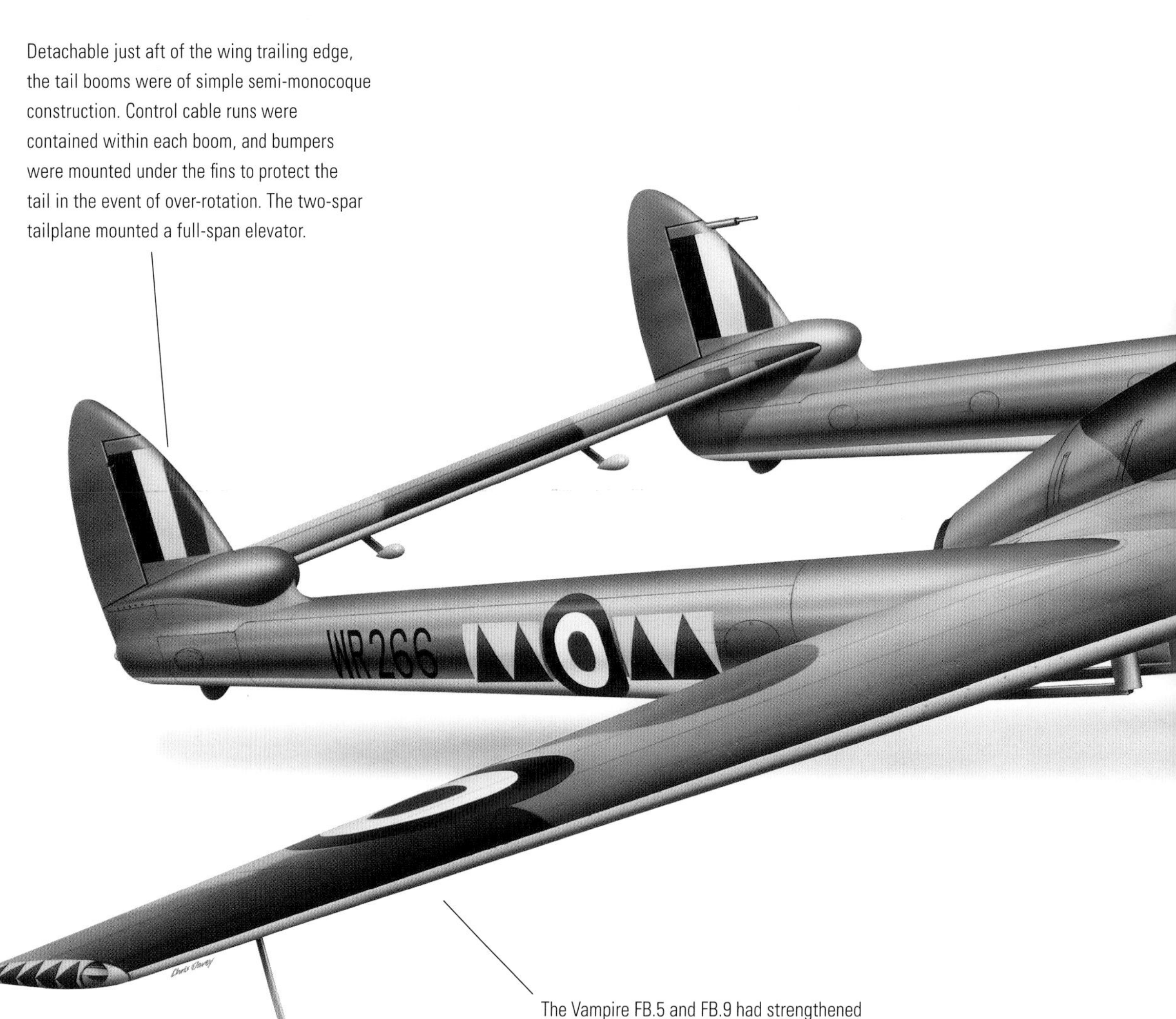

The Vampire FB.5 and FB.9 had strengthened wings of reduced span, allowing underwing stores to be carried. The outer half of the trailing edge was filled by simple ailerons with inboard trim tabs. Small upper and lower airbrake sections were incorporated in the trailing edges next to the ailerons, and inboard of these were two sections of split flaps, one either side of the tail boom.

With no ejection seats fitted to single-seat Vampires, emergency egress was accomplished using the standard bale-out technique. Although the Vampire's cockpit was fairly cramped, its high-set position ensured that the pilot had a superb all-round view.

Based at RAF Ouston in 1956–7, this Vampire FB.9 was operated by No 607 (County of Durham) Squadron, Royal Auxiliary Air Force. The Vampire FB.9 began replacing the FB.5 in April 1956, but the RAuxAF disbanded in March 1957. At that time, ironically, some RAuxAF pilots were already training to fly the swept-wing Hawker Hunter.

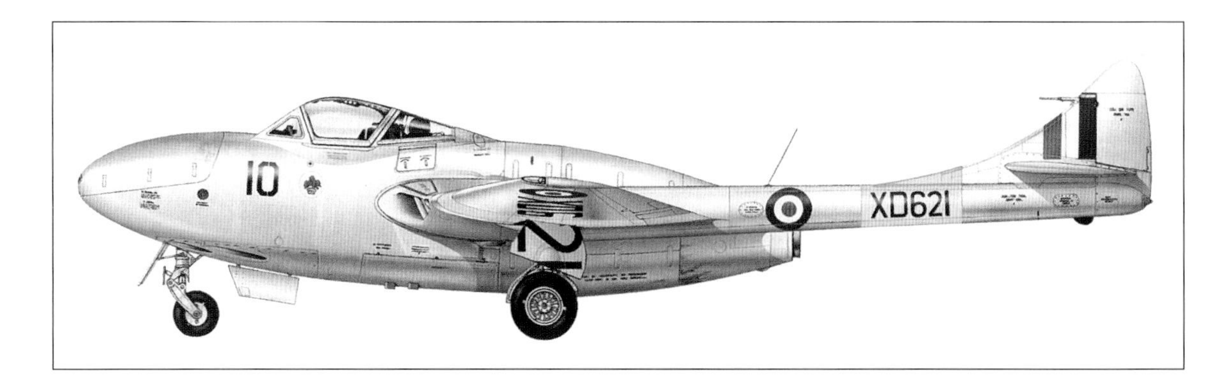

Design work on the DH.100 Vampire, Britain's second jet fighter, began in May 1942, the prototype flying on 20 September 1943, and in the spring of 1944 it became the first Allied jet aircraft capable of sustained speeds of more than 804km/h (500mph) over a wide altitude range. The first production Vampire flew in April 1945 and the Vampire F.1 was delivered to Nos 247, 54 and 72 Squadrons in 1946; 70 aircraft were also delivered to Sweden, some of these being later sold on to the Dominican Republic. After the first 50 aircraft the F.1 was fitted with a pressurized cockpit and a bubble canopy in place of the earlier three-piece hood. The Vampire Mk 2 was a Mk 1 airframe fitted with a Rolls-Royce Nene turbojet and did not enter service: only three were built. It was followed by the Vampire F.3, a long-range version with extra internal fuel, underwing tanks and a de Havilland Goblin 2 turbojet; 85 were supplied to the Royal

Opposite: **Painted in RAF Flying Training Command silver with yellow bands, XD621 was typical of the many Vampire T.Mk.11s which served with eight different flying training schools in the 1950s and 1960s. This example was delivered to No 8 FTS.**

Canadian Air Force, four to Norway and 12 to Mexico, and the type was built under licence in India. The Nene-engined F.Mk.4 was to have been the production version of the Mk.2 and was developed into the F.30/31 built under licence for the Royal Australian Air Force. The first F.30 was delivered to the RAAF on 26 September 1949; 57 were built, followed by 23 FB.31 fighter-bombers, and more than half the F.30s were later modified to FB.31 standard.

The first Vampire variant specifically developed for ground attack was the FB.5, which had squared wingtips, long-stroke naval pattern landing gear and strengthened wings to carry external stores. As the FB.6, fitted with a Goblin 3 engine, it was exported to Switzerland and built under licence in that country, a total of 175 being delivered to the Swiss Air Force. Six were delivered to Finland as the FB.52, being the Finnish Air Force's only post-war combat aircraft until the arrival of 12 Gnat F.Mk.1s in 1958. FB.52s were supplied to the Royal Norwegian Air Force, serving with Nos 336 and 337 Squadrons from 1948; some were supplied to Sweden, replacing the elderly Mk.1s; Mk.5s were supplied to the Royal New Zealand Air Force, equipping No 75 Squadron, and to the Italian and Venezuelan air forces. In December 1949 the Egyptian Air Force received an initial batch of Vampire FB.5s and deliveries continued spasmodically until March 1956, by which time 62 Vampires equipped four first-line squadrons. Vampire FB.5s were also delivered to the South African Air Force from 1950, equipping Nos 1 and 2 Squadrons. The Vampire FB.5 remained in RAF service until 1957, when it was retired on the disbandment of the flying squadrons of the Royal Auxiliary Air Force. The Vampire FB.6 had an uprated Goblin turbojet and was licence-produced in Switzerland. The Vampire FB.9 was a tropicalized version of the FB.5 and was used by the RAF, RNZAF, SAAF, Royal Rhodesian Air Force and India. The Vampire NF.10 was a night fighter, serving with the RAF and the Italian Air Force, in whose service it was designated Mk.54. The Sea Vampire F.20 and F.21 were navalized versions of the Mk.1, the type completing its deck landing trials on the aircraft carrier HMS *Ocean* in December 1945. Only a few were built for carrier trials and jet familiarization. The T.11 was a two-seat trainer. One of the biggest overseas Vampire users was France, which built the Nene-engined Vampire Mk.53 as the Mistral.

Specification: de Havilland Vampire FB.5	
Type:	single-seat fighter-bomber
Powerplant:	one 1420kg (3100lb) thrust de Havilland Goblin 2 turbojet
Performance:	maximum speed 882km/h (548mph) at 9145m (30,000ft); service ceiling 13,410m (44,000ft); range 1960km (1220 miles)
Weights:	empty 3266kg (7200lb); maximum take-off 5600kg (12,348lb)
Dimensions:	wing span 11.58m (38ft 0in); length 9.37m (30ft 9in); height 2.69m (8ft 10in); wing area 24.32m² (262 sq ft)
Armament:	four 20mm (0.79in) British Hispano cannon; up to 907kg (2000lb) of bombs or RPs

Left: **A pair of Swiss Vampires (a T.Mk.55 and an FB.Mk.6 with a modified nose to house reconnaissance cameras) formates with a similarly modified de Havilland Venom, the aircraft that replaced the Vampire on the production line.**

Avro (Hawker Siddeley) Vulcan

The cramped cockpit of the Vulcan normally accommodated a crew of five, although there was provision for two extra crew members. All Black Buck missions were flown by a crew of six, the extra man – either an air refuelling instructor (ARI) or a Vulcan pilot – being carried to assist with refuellings.

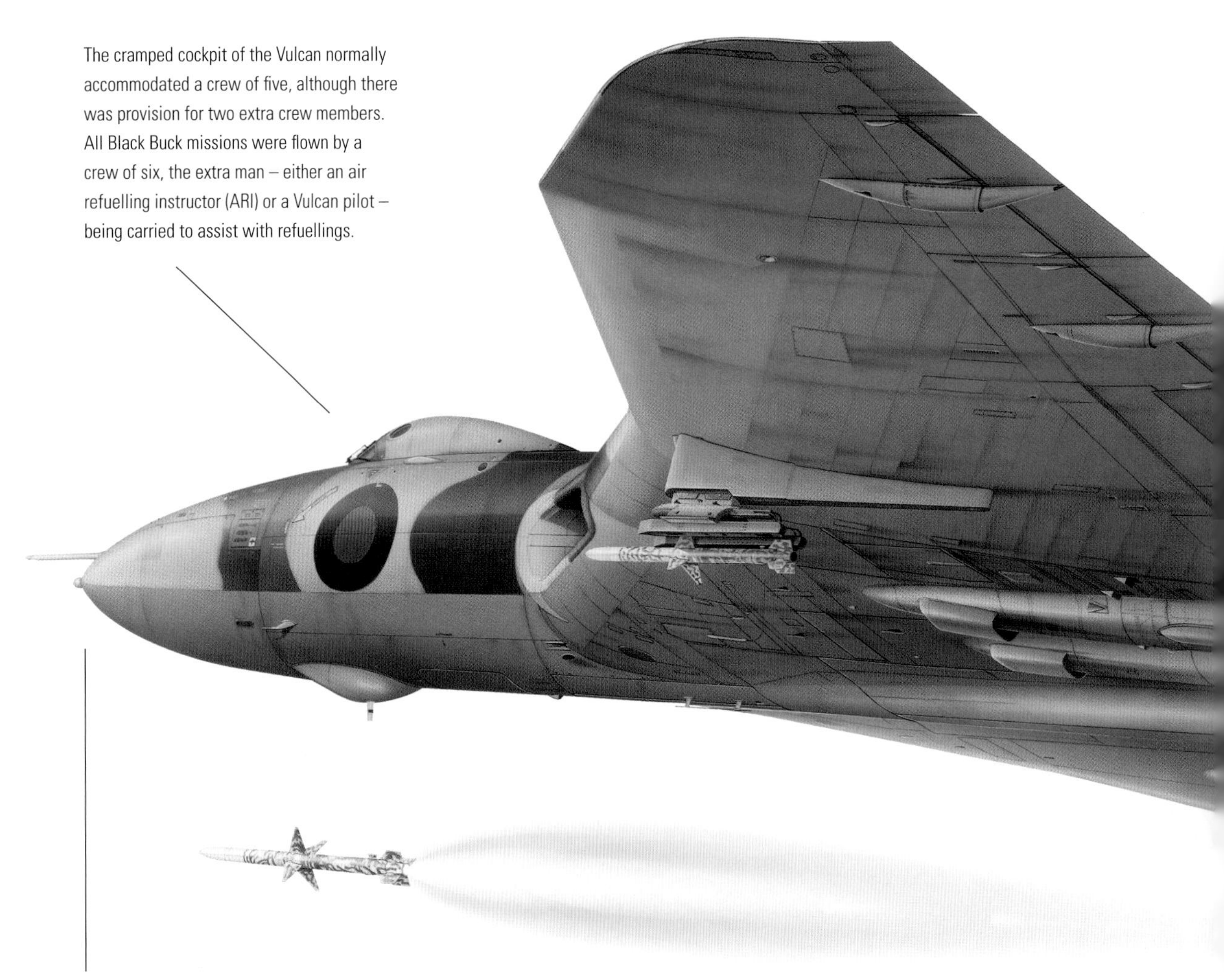

Although designed to carry nuclear weapons, the Vulcan's commodious bomb bay could accommodate 21 x 453kg (1000lb) bombs. The Vulcan's ultimate free-fall nuclear weapon was the WE.177B, designed to be delivered from low level. For this role, the Vulcan was fitted with a terrain-following radar originally designed for the cancelled BAC TSR-2.

This Vulcan B.2, XM597, carried out two anti-radar attacks on the Falklands during the Black Buck missions of May 1982. Flown by Squadron Leader McDougall, it was forced to divert to Rio de Janeiro, Brazil, when its flight refuelling probe fractured after the second sortie. XM597 was later a familiar slight on the display circuit.

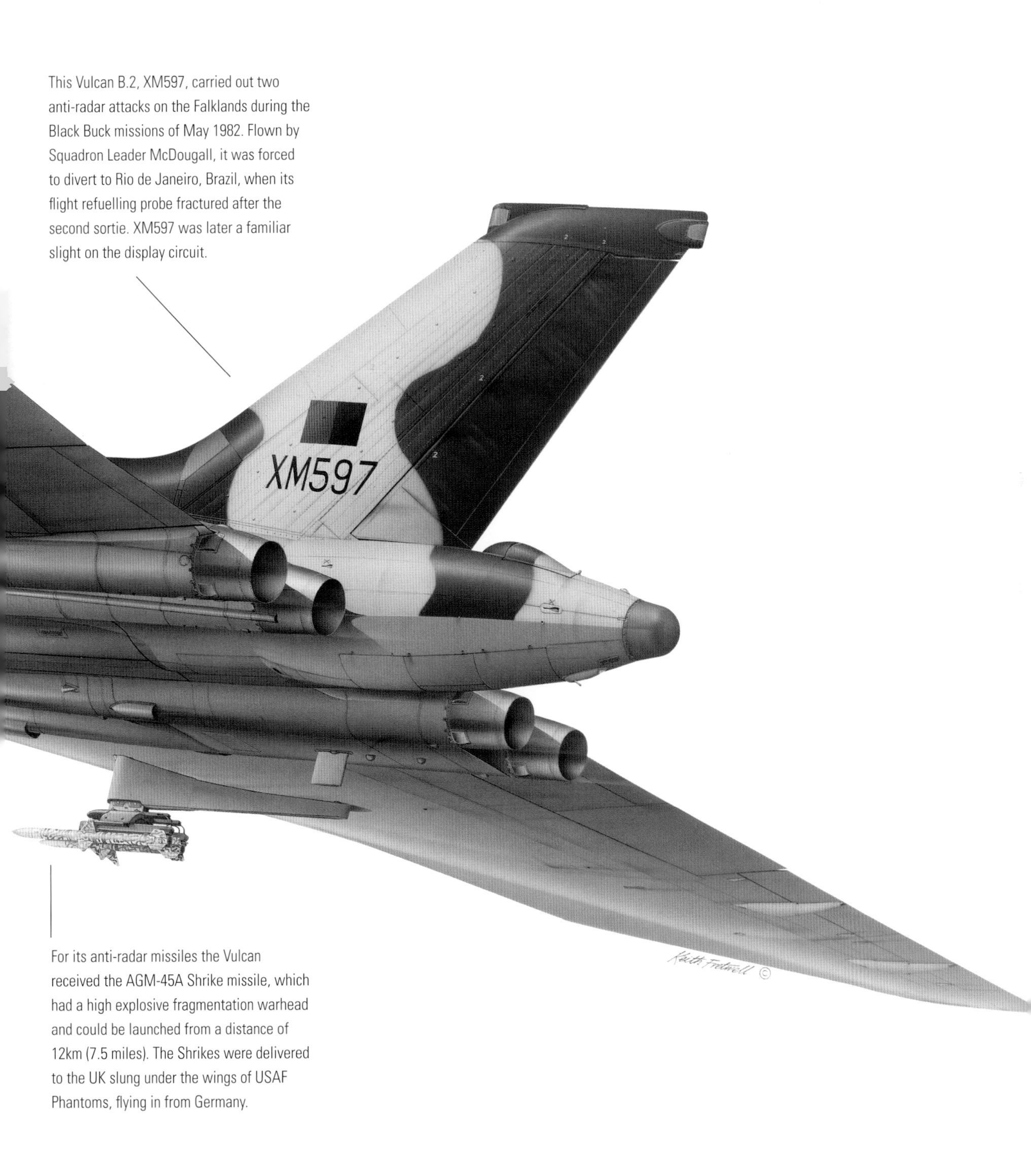

For its anti-radar missiles the Vulcan received the AGM-45A Shrike missile, which had a high explosive fragmentation warhead and could be launched from a distance of 12km (7.5 miles). The Shrikes were delivered to the UK slung under the wings of USAF Phantoms, flying in from Germany.

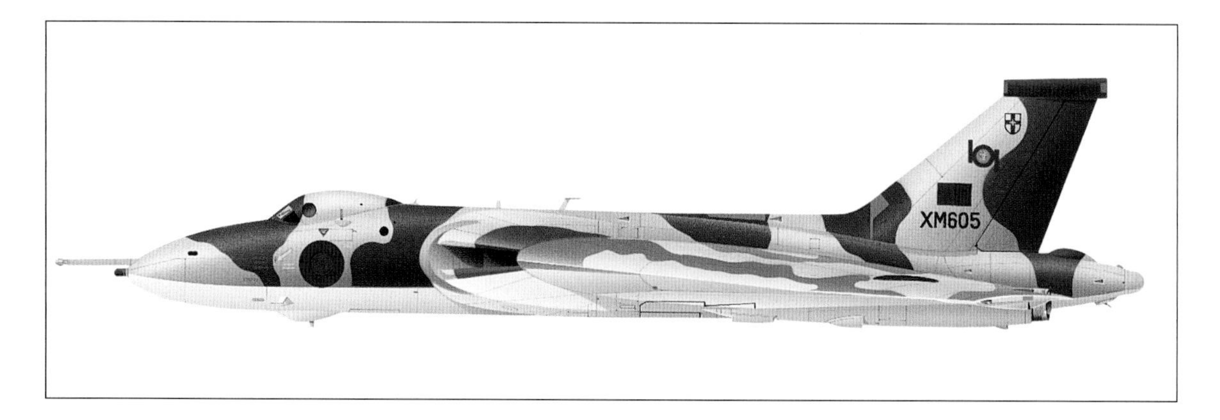

Above: **Delivered in December 1963, Vulcan B.2 XM605 last served with No 101 Squadron, in whose markings it is seen here. In September 1981 it was flown to Castle Air Force Base, California, for permanent display in the USAF Museum.**

The first bomber in the world to employ the delta wing planform, the Avro Type 698 Vulcan prototype (VX770) flew for the first time on 30 August 1952, following extensive testing of its then radical configuration in the Avro 707 series of research deltas. The first prototype was fitted with four Rolls-Royce Avon turbojets and was later re-engined with Bristol Siddeley Sapphires and finally Rolls-Royce Conways, but the second prototype (VX777) employed Bristol Siddeley Olympus 100s. This aircraft, which flew on 3 September 1953, featured a slightly lengthened fuselage and was later fitted with wings having redesigned leading edges with compound sweepback, flying in this configuration on 5 October 1955. It was later used to test the larger wing designed for the Vulcan B.Mk.2, being finally retired in 1960.

The first production Vulcan B.Mk.1 was delivered to No 230 Operational Conversion Unit in July 1956, and No 83 Squadron became the first unit to equip with the new bomber in July 1957. The second squadron to receive the aircraft, in October that year, was No 101, followed in May 1958 by No 617, the famous 'Dam Busters'. By this time production of the greatly improved Vulcan B.Mk.2 was well under way. The first production Vulcan B.2 flew on 30 August 1958, powered by Olympus 200 engines; the second production aircraft featured a bulged tailcone housing electronic countermeasures equipment, and this became standard on subsequent aircraft. Production of the B.Mk.1 was terminated with the 45th aircraft, the remaining Vulcans on order being completed to B.Mk.2 standard with flight refuelling equipment, this model being designed to carry the American Skybolt air-launched IRBM. This was cancelled, but three Vulcan squadrons were armed with the Avro Blue Steel stand-off bomb. Meanwhile, the 34 Vulcan B.Mk.1s remaining in service were withdrawn for conversion to B.Mk.1A standard, which involved the fitting of new avionics, including full ECM. Conversion work was completed early in 1963. Units operating the B.1A were No 44 Squadron, which was formed at Waddington in Lincolnshire on 10 August

1960 by re-numbering No 83; No 50, which re-formed in the following year; and No 101, as well as the Vulcan OCU. The Vulcan B.Mk.2 also received an avionics upgrade, including the fitting of terrain-following radar, and was then designated B.Mk.2A. Having relinquished the QRA (Quick Reaction Alert) role to the Polaris-armed nuclear submarines of the Royal Navy, the RAF's Vulcan force was assigned to NATO and CENTO in the free-fall bombing role. No 27 Squadron's B.2s also operated in the maritime radar reconnaissance role for a time, their aircraft being redesignated Vulcan B.2 (MRR). In May 1982 Vulcans operating from Ascension Island in the Atlantic carried out attacks on the Falkland Islands in support of British operations to recapture them from Argentina.

These operations, code-named 'Black Buck', involved both conventional bombing sorties and anti-radar missions by individual aircraft, each mission being supported by no fewer than 11 sorties by Victor K.2 tankers. Total Vulcan production was 136 aircraft, including the two prototypes and 89 B.2s. The last operational Vulcans were six aircraft of No 50 Squadron, converted for flight refuelling.

Specification: Avro (Hawker Siddeley) Vulcan B.Mk.2	
Type:	five-crew strategic bomber
Powerplant:	four 9072kg (20,000lb) thrust Bristol Siddeley Olympus Mk.301 turbojets
Performance:	maximum speed 1038km/h (645mph) at high altitude; service ceiling 19,810m (65,000ft); range 7403km (4600 miles)
Weight:	maximum take-off 113,398kg (250,000lb)
Dimensions:	wing span 33.83m (111ft); length 30.45m (99ft 11in); height 8.28m (27ft 2in); wing area 368.26m^2 (3964 sq ft)
Armament:	21 x 453kg (1000lb) HE bombs; Yellow Sun Mk.2 or WE.177B nuclear weapons; Blue Steel ASM with Red Snow nuclear warhead (Vulcan B.2BS)

Below: **The most obvious change associated with the V-Force's switch to low-level operations was the adoption of camouflage. Here, a B.2 of the Cottesmore Wing is seen in company with a Vulcan of No 230 OCU, Finningley, finished in white anti-flash gloss.**

Hawker Hunter

The black radome on the nose of the aircraft covers a simple ranging radar. Fighter reconnaissance Hunters replaced this with a forward-facing camera, its lens protected against dirt or insects by automatic 'eyelid' shutters. The small furrow in the upper nose behind the radome serves the camera gun and also acts as a ram air inlet.

The quickly replaceable gun pack fitted to almost all single-seat Hunters contains four 30mm (1.18in) Aden cannon. The gun pack is automatically ventilated, incorporating an electrically operated air scoop that pops out when the guns are fired. Up to 150 rounds of ammunition are provided for each gun, sufficient for 7.4 seconds of firing time.

The Hunter pictured here is an F.Mk.73 of the Sultanate of Oman Air Force. Formerly XG255, a Mk.6 operated by No 66 Squadron RAF, in December 1967 it was brought up to FGA.9 standard for export to the Royal Jordanian Air Force with the designation F.Mk.73A. After eight years of service the aircraft was transferred to Oman in 1975 as an F.Mk.73. The SOAF continued to use the Hunter until 1993.

The outboard wing pylons can accommodate a 455-litre (100 Imp gallons) auxiliary fuel tank, or a range of offensive or defensive weapons. Dutch, Swedish and Swiss Hunters frequently carried AIM-9 Sidewinder short-range AAMs on these pylons.

Early in 1946 both Hawker and Supermarine were studying schemes for swept-wing jet fighters. Two specifications were issued by the Ministry of Supply, both calling for experimental aircraft fitted with swept flying surfaces. Both companies submitted proposals in March 1947, the Hawker design being designated P.1052. This aircraft flew in November 1948, and its performance was such that at one point the air staff seriously considered ordering the type into full production to replace the Gloster Meteor. Instead, the design was developed further under Air Ministry Specification F.3/48, the operational requirement calling for a fighter whose primary role would be the interception of high-altitude, high-speed bombers. The fighter was given the designation P.1067.

The outbreak of the Korean War, together with fears that it might escalate into a wider conflict, led to the acceleration of combat aircraft re-equipment programmes in both east and west. In Britain, the two new swept-wing fighter types, the Hawker P.1067 – soon to be named the Hunter – and Supermarine's design, the Type 541 Swift, flew in prototype form on 20 July and 1 August 1951 respectively and both types were ordered into 'super-priority' production for RAF Fighter Command. The Hunter F.Mk.1, which entered service early in 1954, suffered from engine surge problems during high-altitude gun firing trials, resulting in some

Below: **Hunter FGA.9s of No 208 Squadron at Muharraq, Bahrain, in the early 1960s. The Hunter played a major part in quelling dissident tribes in the troubled Radfan area, and the squadron frequently sent detachments to other Middle East trouble spots.**

modifications to its Rolls-Royce Avon turbojet, and this – together with increased fuel capacity and provision for underwing tanks – led to the Hunter F.4, which gradually replaced the Canadair-built F-86E Sabre (which had been supplied to the RAF as an interim fighter) in the German-based squadrons of the 2nd Tactical Air Force. The Hunter Mks 2 and 5 were variants powered by the Armstrong Siddeley Sapphire engine. In 1953 Hawker equipped the Hunter with the large 10,000lb thrust Avon 203 engine, and this variant, designated Hunter F.Mk.6, flew for the first time in January 1954. Deliveries began in 1956 and the F.6 subsequently equipped 15 squadrons of RAF Fighter Command. The Hunter FGA.9 was a development of the F.6 optimized for ground attack, as its designation implies. The Hunter Mks 7, 8, 12, T52, T62, T66, T67 and T69 were all two-seat trainer variants, while the FR.10 was a fighter-reconnaissance version, converted from the F.6. The GA.11 was an operational trainer for the Royal Navy.

In a career spanning a quarter of a century the Hunter equipped 30 RAF fighter squadrons, in addition to numerous units of foreign air forces. The aircraft was licence-built in the Netherlands and Belgium; principal customers for British-built aircraft were India, Switzerland and Sweden. Indian Hunters saw considerable action in the 1965 and 1971 conflicts with Pakistan, ten Hunters being lost in the three-week air war of 1965 and 22 in the 1971 battle, some of these being destroyed on the ground. The grand total of Hunter production, including two-seat trainers, was 1972 aircraft, and over 500 were subsequently rebuilt for sale overseas.

Below: **A Hunter T.Mk.8 of No 738 Naval Air Squadron, based at RNAS Brawdy during 1965, XL598 was one of the first batch of ten new-build T.8s (originally ordered as T.7s for the RAF) and made its first flight on 15 October 1958. The aircraft is marked with Brawdy's 'BY' codes and the Pegasus from its squadron emblem.**

In August 1953 the prototype Hawker P.1067 was fitted with a Rolls-Royce RA.7R afterburning engine, in effect a 'racing' Avon, for an attack on the World Absolute Air Speed Record. Fitted with a sharply pointed nosecone fairing, the aircraft was flown to Tangmere, Sussex, at the end of August for practice runs. On 7 September, Hawker's Chief Test Pilot, Neville Duke, broke the record with an average speed of 1171km/h (727.63mph). Twelve days later, the aircraft also established a 100km closed circuit world record at an average speed of 1141km/h (709.2mph). The aircraft, subsequently referred to as the Hunter Mk.3, is now part of the RAF Museum collection.

The success of the Hunter airframe and afterburning engine combination led to a proposal for a supersonic Hunter variant, the Hawker P.1083. Construction of a prototype was begun, but the project was cancelled when this was more than 80 per cent complete.

Specification: Hawker Hunter F.Mk.6	
Type:	single-seat fighter-bomber
Powerplant:	one 4535kg (10,000lb) thrust Rolls-Royce Avon 203 turbojet
Performance:	maximum speed 1117km/h (694mph) at sea level; service ceiling 14,325m (47,000ft); range 689km (428 miles)
Weights:	empty 6406kg (14,122lb); maximum take-off 7802kg (17,200lb)
Dimensions:	wing span 10.26m (33ft 8in); length 13.98m (45ft 10in); height 4.02m (13ft 2in); wing area 32.42m² (349 sq ft)
Armament:	four 30mm (1.18in) Aden cannon; underwing pylons with provision for two 453kg (1000lb) bombs and 24 x 76mm (3in) rockets

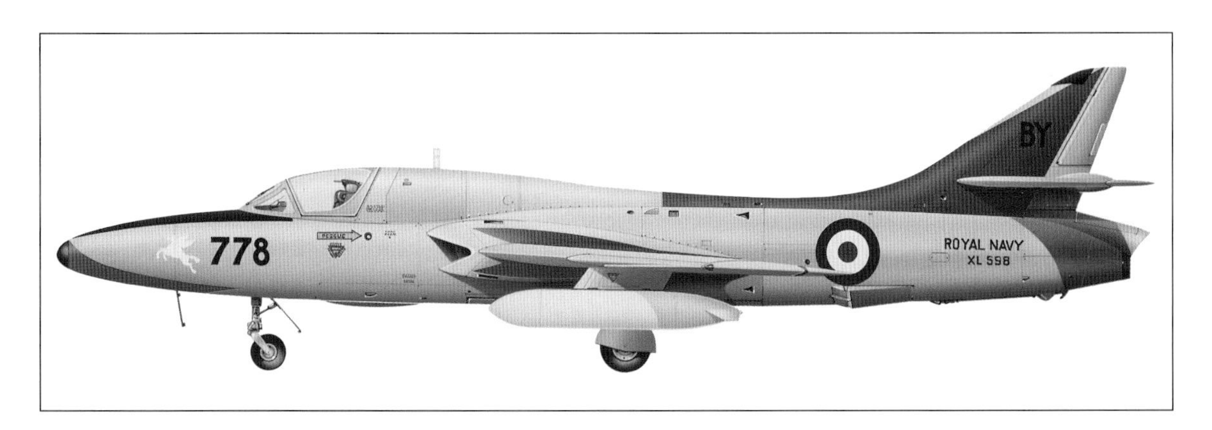

Handley Page Victor

Only a small portion of the Victor's high-mounted tailplane was fixed, the majority of it moving to act as a giant elevator. The bullet fairing that covered the fin/tailplane joint housed an antenna at either end for the ARI 18228 radar warning receiver. At the base of the fin was a large intake for a heat exchanger, added to the B.Mk.2 to cool the ECM equipment.

Mounted under each wing was a Flight Refuelling Ltd FR.20B hose-drogue unit, which could contain 659 litres (145 Imp gallons) of fuel. The Mk.20 was equipped with a 15m (49ft) hose, which deployed by air resistance. It was winched in by power from the ram air turbine on the front of the pod.

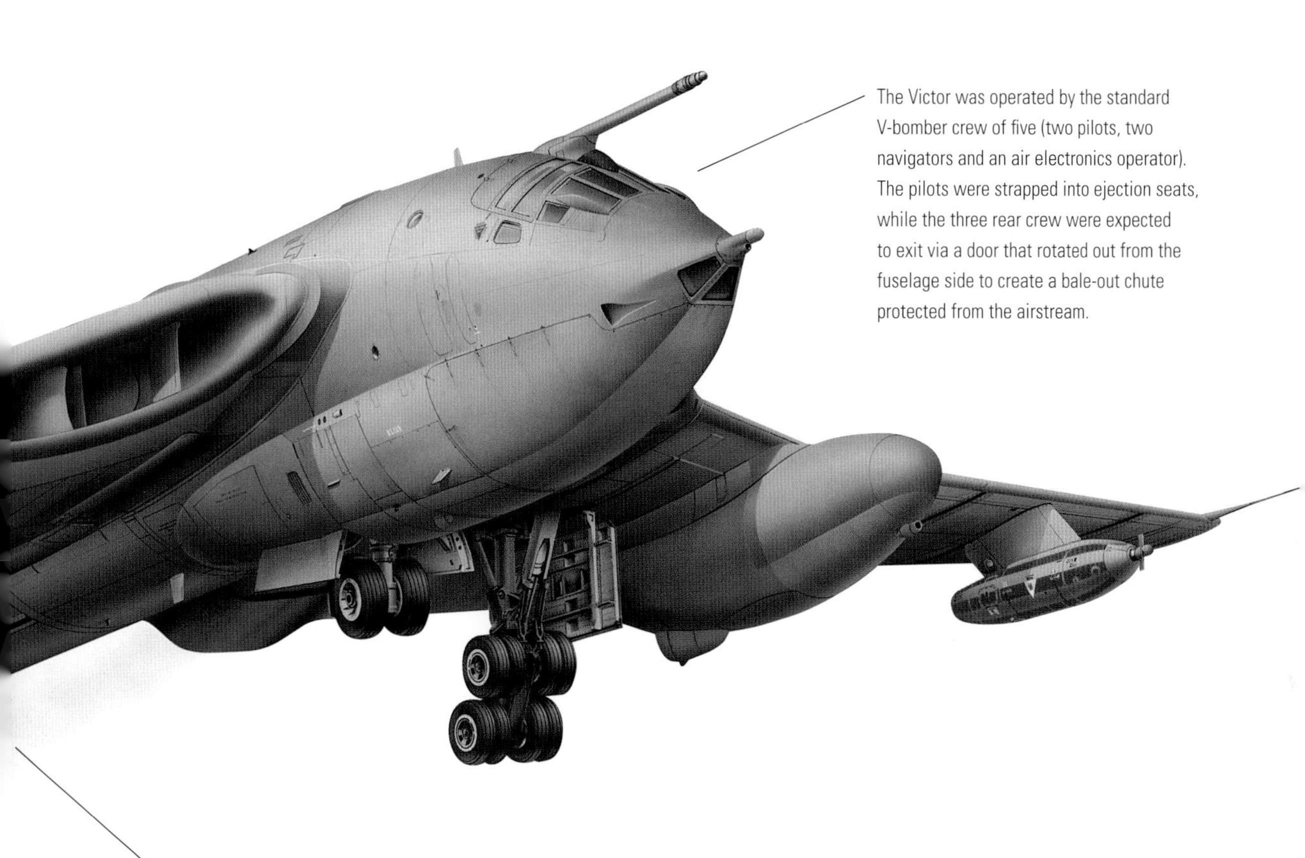

The Victor was operated by the standard V-bomber crew of five (two pilots, two navigators and an air electronics operator). The pilots were strapped into ejection seats, while the three rear crew were expected to exit via a door that rotated out from the fuselage side to create a bale-out chute protected from the airstream.

Mounted in the rear of the former bomb bay, the Flight Refuelling Ltd Mk 17 HDU could deploy up to 24.70m (81ft) of hose, and its flow rate of up to 1814kg (4000lb) or 2273 litres (500 Imp gallons) per minute was more suitable for refuelling large aircraft. Furthermore, the drogue was more stable in the airflow than those deployed from the wing pods.

The last in a long line of Handley Page bombers, and the last in the RAF's trio of V-bombers, the HP.80 Victor's design owed much to research into the crescent wing carried out by the German Arado and Blohm & Voss firms in World War II. The prototype HP.80 Victor, WB771, was flown from Boscombe Down on 24 December 1952 by Handley Page's Chief Test Pilot, Sqn Ldr H.G. Hazelden, with E.N.K. Bennett as his flight observer. The maiden flight was effortless, and it was during the landing that the Victor displayed one of its finest handling characteristics: if set up properly on final approach it would practically land itself. When most aircraft enter the ground cushion in the round-out stage, just before touchdown, the ground effect tends to destroy the downwash from the tailplane, causing a nose-down moment and making it necessary for the pilot to hold off with backward pressure on the control column; the Victor's high-set tailplane eliminated this effect almost entirely. Also, the aircraft's crescent wing configuration reduced downwash at the root and upwash at the tips, a characteristic of normal swept wings, and this produced a nose-up pitch that contributed to a correct landing attitude.

The HP.80's structure was every bit as remarkable as its aerodynamics, and incorporated many features that were radical departures from any previous techniques. The wing, basically, was of multi-spar construction with load-carrying skins forming multiple torsion boxes; the inner part of each wing was of three-spar construction, with a four-spar structure outboard of the landing gear. The all-metal ailerons operated through Hobson electrically actuated, hydraulically powered control units; there were hydraulically operated Fowler flaps on the inboard trailing edges, with two-piece hydraulically actuated leading-edge flaps on each outer wing (these were later replaced by 'drooped' leading edges). The wing itself was of sandwich construction, with a corrugated core of aluminium alloy sheet for the skin and components such as ribs and spar webs, resulting in considerable strength combined with weight saving. The wing box ran ahead of the engines, which were completely buried in the thick inboard section of the wings; there was adequate room for the installation of larger and more powerful units, and less risk of damage

Below: **A Handley Page Victor B.2(BS) of No 139 Squadron, RAF Wittering. No 139 was one of two Victor Blue Steel squadrons, the other being No 100, both based at RAF Wittering near Peterborough. The Blue Steel missile is partly recessed in the bomb bay.**

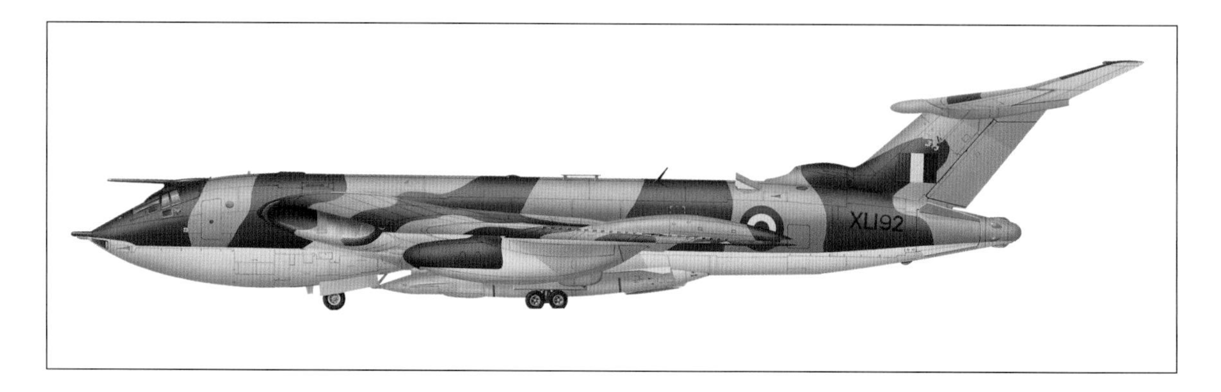

Above: **The second production Victor B.Mk.1, XA 918, flew in March 1965 and served as a trials aircraft throughout its career. As one of the first four B.Mk.1s, it initially sported an overall silver finish, later changed to white as a protection against the effects of nuclear flash.**

to the wing box if an engine caught fire or a turbine disintegrated. Much use was made of spot welding in attaching the outer skin to the core of the wing, a very bold move on the part of Handley Page at that time.

The Victor prototype was destroyed when the tailplane broke away during a low-level run. The second prototype flew on 11 September 1954, followed by the first production Victor B.Mk.1 on 1 February 1956. The first Victor squadron, No 10, became operational in April 1958, and three more, Nos 15, 55 and 57, had formed by 1960. The B.Mk.1A was an updated variant with more advanced equipment, including ECM in the tail, and the B.Mk.2 was a more powerful version with a larger span. The B.Mk.2 was designed to carry the cancelled US Skybolt IRBM, and two squadrons (Nos 100 and 139) were armed with the Avro Blue Steel stand-off missile. The Victor B.(PR).Mk.1 and B.(PR).Mk.2 were photo-reconnaissance variants, both serving with No 543 Squadron. In 1964–5 the earlier Victors were converted to the flight refuelling tanker role as B.(K).Mk 1s and 1As, and 27 Mk.2s were converted to K.Mk.2 tankers in 1973–4. These aircraft served with Nos 55 and 57 Squadrons and were withdrawn from service in the early 1990s, after participating in the Gulf War. Victor production totalled 50 B.1s/1As and 34 B.2s.

Specification: Handley Page Victor B.Mk.2(BS)

Type:	five-crew strategic bomber/tanker
Powerplant:	four 9344kg (20,600lb) thrust Rolls-Royce Conway Mk.201 turbofans
Performance:	maximum speed 1040km/h (645mph) at 12,190m (40,000ft); service ceiling 14,325m (47,000ft); range 7400km (4600 miles)
Weights:	empty 41,268kg (91,000lb); maximum take-off 105,665kg (233,000lb)
Dimensions:	wing span 36.58m (120ft 0in); length 35.05m (115ft); height 9.20m (30ft 2in); wing area 223.52m² (2406 sq ft)
Armament:	one HS Blue Steel ASM (Red Snow warhead)

English Electric Lightning

The Lightning F.Mk.3 seen here, XP762, wears the insignia of No 111 (Treble One) Squadron, which was based at RAF Wattisham in Suffolk. The unit re-equipped with the Lightning F.Mk.1A in 1961, subsequently converting to the F.Mk.3 in 1964. Treble One continued to operate the Lightning until 1974, when it received the Phantom FG.1/FGR.2.

The nose section was built in two halves to allow for easy access for the installation of wiring, pneumatic and hydraulic lines. The windscreen and canopy frames were manufactured from heavy forgings, with stretched acrylic Perspex mouldings and an optically flat armoured glass windscreen. After final assembly, the cockpit was pressure tested to two and a half times the maximum pressure expected in service. The nose section was then joined, at the aft cabin pressure bulkhead, to the main fuselage.

Developed by de Havilland Propellers, the Firestreak AAM was first deployed on the Royal Navy's Sea Venoms. It was an infrared system, designed to lock on to the heat emissions from the target aircraft's engine exhaust.

Apart from the Ferranti AIRPASS radar, the Lightning was fitted with TACAN for air navigation. From the F.Mk.2 onwards the aircraft could use an Offset TACAN display, allowing the pilot to 'move' a TACAN beacon to a location of his choice, thereby allowing him to make instrument let-downs at non-TACAN airfields.

Above: **No RAF two-seaters were built with the larger ventral fuel/weapons pack, unlike the Saudi T.Mk 55, which might have formed the basis of a truly capable multi-role warplane. The aircraft illustrated is a Lightning T.Mk 5.**

Only the RAF, of all the world's air forces, made the jump from subsonic to Mach 2 fighter with no Mach 1 plus intermediary, replacing the Hawker Hunter day fighter and the Gloster Javelin all-weather fighter with the Mach 2 English Electric (later BAC) Lightning. The English Electric Lightning was based on the P.1A research aircraft, which first flew on 4 August 1954, powered by two Bristol Siddeley Sapphires. Three operational prototypes, designated P.1B, were also built. The first of these flew on 4 April 1957, powered by two Rolls-Royce Avons, and exceeded Mach 1.0 on its first flight. On 25 November 1958 it became the first British aircraft to reach Mach 2.0, which it did in level flight. By this time the P.1B had been given the name Lightning and ordered into production for RAF Fighter Command.

The first production Lightning F.Mk.1 flew on 29 October 1959, and fully combat-equipped Lightnings began entering RAF service in July 1960. The Lightning, which had a phenomenal initial climb rate of 15,240m (50,000ft) per minute, was constantly improved during its career, evolving via the F.2 and F.3 into the F.6 version. This had a revised wing leading edge designed to reduce subsonic drag and improve range, and was fitted with a large ventral fuel pack with more than double the capacity of earlier packs. The first Lightning F.6 flew in April 1964 and entered service in the following year. It was the last jet fighter of purely British design, and it was to serve the RAF well in the front line of NATO's air defences until its eventual retirement in 1976. Lightnings were also supplied to the Saudi Arabian and Kuwait air forces as the Mk.53/Mk.55.

Opposite: **A two-seat Lightning T.5 of No 5 Squadron leads a pair of F.6s of No 11 Squadron from RAF Binbrook, Lincolnshire. Binbrook was the last operational Lightning station in the UK. Nos 5 and 11 Squadrons disbanded in 1987–88 and reformed immediately with Tornado F.3s.**

All Lightnings shared the same engine layout, which gave the type its characteristic slab-sided appearance. The two engines were mounted one above the other and staggered, with the lower engine forward. This gave the smallest possible fuselage cross-section, but necessitated the use of different-length jet pipes, and made the engines vulnerable to collateral damage if one failed. Both engines received air from a nose intake, via complex, horizontally bifurcated internal ducts. The top engine was removed upwards (after removal of the jet pipe, which rolled out aft) while the lower engine was dropped down. An engine change could theoretically be achieved in four hours, although in practice it usually took days. Complex and inaccessible systems on the aircraft, coupled with a poor logistics system, meant that the early Lightnings had a poor utilization rate. This was greatly improved with later marks of Lightning, but the systems were far from popular with engineers and ground crews. From the aircrew point of view, the Lightning was a true fighter pilot's aircraft, demanding and difficult to master, yet ultimately highly rewarding.

In order to achieve performance above Mach 2, the Lightning's wings were swept back sharply. In planform, the wings resembled a notched delta with a leading edge sweep angle of 60 degrees to ensure that the vortex flow would be fully developed. To ensure maximum effectiveness, the ailerons were mounted normal to the airflow on the cut-off tips. The wing also had trailing-edge flaps and (on production aircraft) a fixed but detachable leading edge. The latter was entirely clean except for a small 'saw-cut notch' at around two-thirds span. This was added to cure uneven airflow around the ailerons, a problem encountered during low-speed flight testing of the Short SB.5 research aircraft, built specially to test the wing configuration. The notch was fitted in place of a wing fence, which would have created higher drag, and was found to have the additional beneficial effect of re-energizing spanwise airflow, and thus increasing vortex-derived lift. The wing was joined at the aircraft centreline and was installed as a one-piece unit, making removal of the wings difficult and time-consuming after manufacture was complete. The wing was built around two primary spars, to which were affixed closely spaced ribs and stringers. The wing torsion boxes formed the aircraft's primary internal fuel tankage, this being restricted by the

space taken up by the main undercarriage wells. Shortage of fuel was a serious problem with the early marks of Lightning, later redressed to a great extent by the fitting of a large ventral fuel tank and flight refuelling equipment. Fuel consumption, however, remained extremely high: at low level, at 1110km/h (690mph) each Avon engine devoured 91kg (200lb) of fuel per minute, giving a typical combat air patrol (CAP) time of only 12 minutes. The pilot had to return to base with a minimum of 726kg (1600lb) of fuel remaining, to allow for diversions or missed approaches.

The standard Lightning intercept aimed to approach the target on a reciprocal (head-on) course and slightly below, so that the radar looked up, displaced laterally by 13km (8 miles), with the Lightning turning in to roll out behind the target at about 2.4km (1.5 miles). Conveniently, the Lightning's turn radius at Mach 0.85 and 7620m (25,000ft) was 6.5km (4 miles). In such an intercept, the pilot had to remember that the blip should be 20 degrees off the nose at 40km (25 miles), 25 degrees at 32km (20 miles), 32 degrees at 24km (15 miles) and 40 degrees at the turn point (19km/12 miles). Turn radii varied with height and speed, which meant more curves for the pilot to memorize. As if this were not difficult enough, the Lightning pilot had to calculate the target's inbound heading by measuring change of displacement over 8km (5 miles) of closure, and calculated the target height at 16km (10 miles) by multiplying scanner elevation by 10. This was vital, since the height difference between target and fighter had to be reduced to within 609m (2000ft) to stand any chance of success, preferably with the fighter just below the target. Locking on to the target, the radar provided range and azimuth information, but no closure rate. If a missile was fired, the Lightning then had to roll inverted and pull inverted, in order to avoid debris from the target. This procedure served to underline the Lightning's limitations as an interceptor. Nevertheless, the Lightning made a tremendous contribution to UK air defence for many years.

Left: **A Lightning F.6 of No 23 Squadron, RAF Leuchars, Scotland, intercepts a Soviet Tu-95 Bear electronic intelligence aircraft over the North Sea. Such interceptions were frequent until the collapse of the Soviet Union and the end of the Cold War.**

Specification: English Electric Lightning F.Mk.3	
Type:	single-seat interceptor
Powerplant:	two 7112kg (15,680lb) thrust Rolls-Royce Avon 211R turbojets
Performance:	maximum speed 2415km/h (1500mph, Mach 2.3) at 12,190m (40,000ft); service ceiling 18,920m (62,000ft) plus; range 1287km (800 miles)
Weights:	empty 12,700kg (28,000lb); maximum take-off 22,680kg (50,000lb)
Dimensions:	wing span 10.61m (34ft 10in); length 16.84m (55ft 3in); height 5.97m (19ft 7in); wing area 35.31m^2 (380.1 sq ft)
Armament:	two nose-mounted 30mm (1.18in) Aden guns; two Firestreak or Red Top AAMs

British Aerospace (Hawker Siddeley) Hawk

Both crew members have a zero-zero (zero altitude, zero speed) escape capability provided by a Martin-Baker Mk 10B rocket-assisted ejection seat. Instructors can automatically eject themselves first, and then the student within 0.55 seconds, using the command ejection system. Alternatively, each occupant can leave independently by using the seat pan handle.

The Hawk has conventional hydraulically actuated ailerons and one-piece all-moving tailplanes, with a mechanically actuated rudder. The double-slotted trailing-edge flaps inboard are also hydraulically actuated, as is the large airbrake under the rear fuselage, between the fixed ventral fins.

This Hawk T.Mk.1A is seen in the colours of No 92 (Reserve) Squadron, No 7 Flying Training School, which was based at RAF Chivenor in Devon until its closure on 1 October 1994. With the disbandment of No 92 Squadron, this Hawk, XX157, went on to serve with the Royal Navy's Fleet Requirements and Air Direction Unit (FRADU) at RNAS Yeovilton. No 92 Squadron's badge depicts a cobra (the squadron was raised with funds gifted by India and bears the title 'East India') and a maple sprig, recalling its association with Canada in World War I, when most of its personnel were Canadian.

A detachable pod beneath the fuselage contains a single Aden Mk 4 cannon; this can be loaded with training ammunition (ball, Practice Round Mk 4), high explosive (Mk 6) or armour-piercing (Mk 1) shells. Rate of fire is 1200–1400 rounds per minute and the pod has a detachable rear section to allow for the rapid insertion of the 130-round magazine during operational turnrounds.

Designed as a Gnat and Hunter replacement in the advanced training and strike roles, the Hawker Siddeley Hawk prototype flew in August 1974 and the first two operational Hawk T.Mk.1s (from an eventual total of 175) were handed over in November 1976. The Hawk T.Mk.1A is a tactical weapons trainer with three pylons, the one on the centreline normally occupied by a 30mm (1.18in) Aden gun pack. The Hawk Series 60 and Series 100 are two-seat export versions, while the Hawk 200 is a single-seat dedicated ground attack variant, designed from the outset to be a cost-effective, multi-role combat aircraft. Hawks have been exported to some 20 countries, and a modified variant serves with the US Navy as the T-45A Goshawk. Hawks in RAF service also provide target facilities for front-line fighter units, having replaced the venerable Canberra in that role.

Below: **The British Aerospace Hawk demonstrator, bearing the civilian registration G-HAWK, pictured in company with another demonstration aircraft, British Aerospace Harrier T. Mk.52 G-VTOL. The Hawk also carried the military serial ZA101.**

One interesting potential use for the Hawk during the later years of the Cold War was its assignment to a concept known as the Mixed Fighter Force, which involved missile-armed Hawks forming a second line of defence while Tornado ADVs and Phantoms formed the first line. The

aim of the Hawk War Role Programme, as it was known, was to provide 89 Hawks, whose sole armament then (in 1979) was a 30mm (1.18in) Aden gun, with an air defence capability by fitting two Sidewinder AAMs to each aircraft. In January 1983, following successful trials, British Aerospace received a contract to modify the 89 Hawks to War Role standard for the secondary defence of UK installations. Aircraft so modified – those in service with Nos 1 and 2 Tactical Weapons Units at RAF Brawdy and RAF Chivenor, together with those of the Central Flying School at Scampton (including the aircraft used by the Red Arrows, the RAF aerobatic team) – were allocated the designation Hawk T.Mk.1A.

The modification programme was completed in August 1986, by which time the original concept of using Hawks for point defence had undergone changes as a result of lessons learned during air defence exercises. The Hawk's original task of acting independently on a day visual CAP (combat air patrol) was now reduced to a secondary role; the primary role was to operate in concert with the longer-range Tornado F.3s and Phantoms, the idea being to catch incoming enemy aircraft as far out as possible from the British coastline. Each Phantom or Tornado would fly in company with one or two Hawks in loose formation, waiting to be advised of a threat by AWACS, an air defence vessel, ground radar or the aircraft's own AI radar. The Tornado or Phantom would then use its radar to set up the engagement and, because of the threat of severe jamming, would use visual signals to alert the Hawk. As soon as radar contact was established the pilot would stand his aircraft on a wingtip and then roll it around its axis to stand on the other; the Hawk pilot would reply in similar fashion

Above: **The Red Arrows RAF aerobatic team, famous throughout the world for their precision display routine, have used the B.Ae Hawk since 1980. Before that the team used the Hawker Siddeley Gnat, which the Hawk replaced as the RAF's advanced jet trainer.**

by way of acknowledgement. Depending on the Rules of Engagement in force at the time, the Tornado/Phantom might be in a position to launch its long-range missiles, in which case the first clue the Hawk pilot might have that a hostile had been detected was when he saw the smoke trails. If the incoming hostiles were bombers with a fighter escort, the Hawks would take on the fighters while the more heavily armed aircraft went for the bombers.

Specification: British Aerospace Hawk T.Mk.1A

Type:	two-seat tactical weapons trainer
Crew:	2
Powerplant:	one 2359kg (5200lb) thrust Rolls Royce/Turboméca Adour Mk 151 turbofan
Performance:	maximum speed 1038km/h (645mph); service ceiling 15,240m (50,000ft); endurance 4 hours
Weights:	empty 3647kg (8040lb); maximum take-off 7750kg (17,085lb)
Dimensions:	wing span 9.39m (30ft 9in); length 11.17m (36ft 7in); height 3.99m (13ft 1in); wing area 16.69m² (179.64 sq ft)
Armament:	under-fuselage/wing hardpoints for up to 2567kg (5660lb) of stores; wingtip-mounted AAMs; one Aden Mk 4 cannon in detachable pod

British Aerospace Sea Harrier

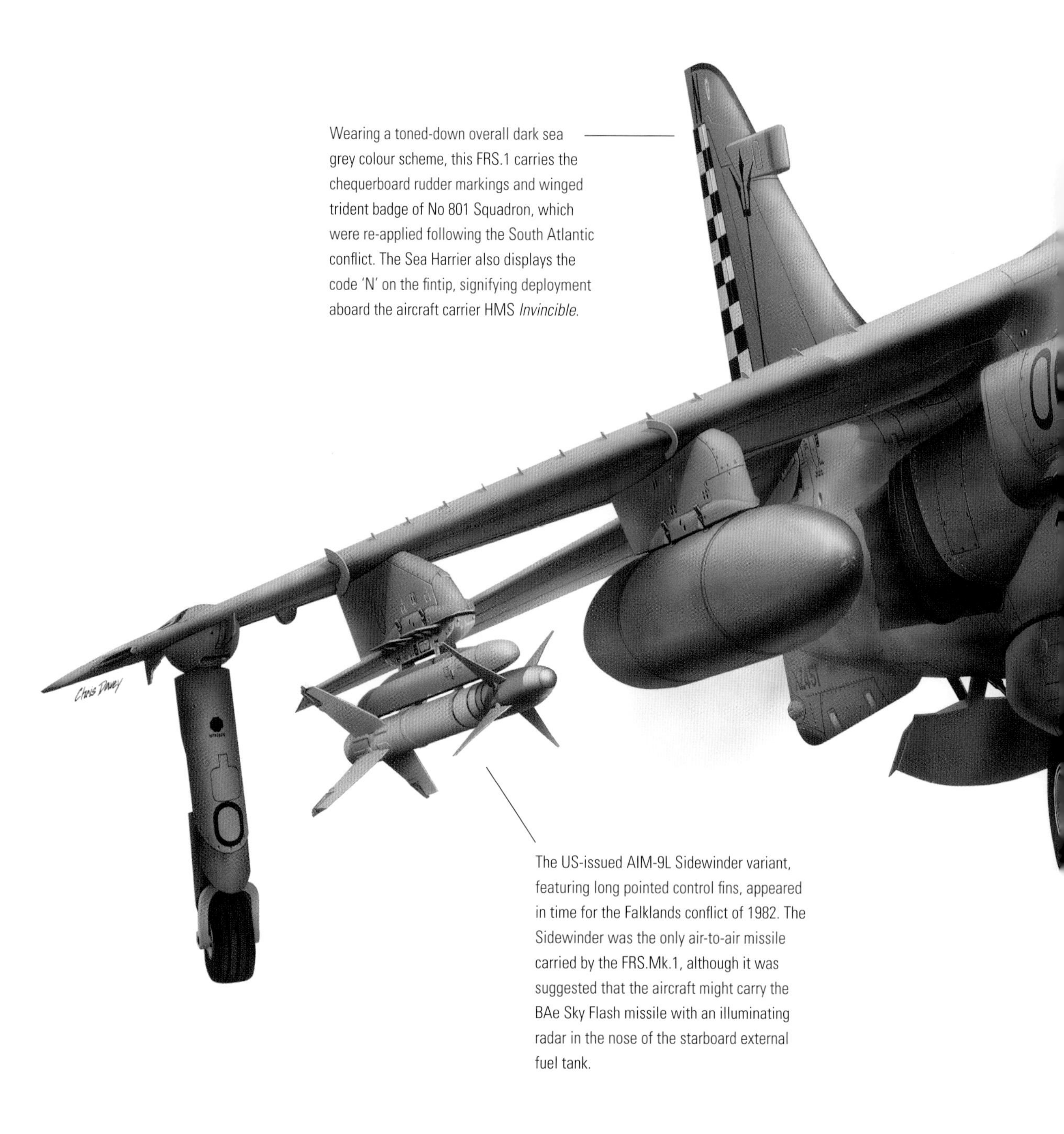

Wearing a toned-down overall dark sea grey colour scheme, this FRS.1 carries the chequerboard rudder markings and winged trident badge of No 801 Squadron, which were re-applied following the South Atlantic conflict. The Sea Harrier also displays the code 'N' on the fintip, signifying deployment aboard the aircraft carrier HMS *Invincible.*

The US-issued AIM-9L Sidewinder variant, featuring long pointed control fins, appeared in time for the Falklands conflict of 1982. The Sidewinder was the only air-to-air missile carried by the FRS.Mk.1, although it was suggested that the aircraft might carry the BAe Sky Flash missile with an illuminating radar in the nose of the starboard external fuel tank.

Between the twin gun pods, the characteristic Dowty Rotol twin-wheel main landing gear supports almost all the aircraft's weight when on deck, the outrigger wheels having a mainly stabilizing function. For landing on carrier decks, the unit is fitted with adaptive anti-skid brakes.

The Sea Harrier FRS.1 was armed with two 30mm (1.18in) Aden cannon in pods under the fuselage, each gun with 150 rounds of ammunition. In the Sea Harrier FA.2, two 25mm (1in) Aden guns are carried.

Developed from the basic Harrier airframe, the Sea Harrier FRS.1 was ordered to equip the Royal Navy's three Invincible-class aircraft carriers. The nose was lengthened to accommodate the Blue Fox AI radar, and the cockpit was raised to permit the installation of a more substantial avionics suite and to provide the pilot with a better all-round view. The first Sea Harrier FRS.1 took off for its maiden flight from Dunsfold on 20 August 1978; this aircraft, XZ450, was not in fact a prototype but the first aircraft of a production order that had now risen from 24 to 31. On 13 November it became the first Sea Harrier to land on an aircraft carrier, HMS *Hermes*. In addition to the production batch, three development Sea Harriers had been ordered in 1975. The first of these, XZ438, flew on 30 December 1978 and was retained by the manufacturers for performance and handling trials. The second, XZ439, flew on 30 March 1979 and went to the A&AEE Boscombe Down for stores clearance trials, while the third, XZ440, flew on 6 June 1979 and was employed in handling and performance trials at BAe Dunsfold, Boscombe Down, and with the RAE and Rolls-Royce (Bristol).

The second production Sea Harrier, XZ451, flew on 25 May 1979 and became the first example to be taken on charge by the Royal Navy, being accepted on 18 June 1979 for service with the Intensive Flying Trials Unit. No 800A Naval Air Squadron was commissioned at Royal Naval Air Station, Yeovilton, Somerset, on 26 June 1979 as the Sea Harrier Intensive Flying Trials Unit (IFTU), and on 31 March 1980 this unit was disbanded and reformed as

Below: **A Royal Navy Sea Harrier FA.2 lands on a carrier deck. The FA.2 incorporated a number of important upgrades, including a redesign of the front fuselage to accommodate the Ferranti Blue Vixen pulse-Doppler radar.**

Above: **The B.Ae Sea Harrier FA.2 was the final version of the type, with improved radar and armament. After the Sea Harrier's withdrawal, fleet protection will be the task of the RAF's Harrier GR.7s, flown by RAF and Royal Navy pilots.**

No 899 Headquarters and training squadron. A second Sea Harrier squadron, No 800, was commissioned on 23 April 1980, and was followed by No 801 Squadron on 26 February 1981. The planned peacetime establishment of each squadron was five Sea Harriers; No 800 was to embark on HMS *Hermes*, while No 801 was to go to HMS *Invincible*. Meanwhile, an additional batch of ten Sea Harriers had been ordered from British Aerospace; the first of these flew on 15 September 1981 and was delivered to No 899 Squadron.

Armed with Sidewinder AAMs, the Sea Harrier FRS.1 distinguished itself in the 1982 Falklands War. At the height of the campaign, on 21 May 1982, Sea Harriers were being launched on combat air patrols at the rate of one pair every 20 minutes. The Sea Harrier force was later upgraded to FA.2 standard.

Specification: Sea Harrier FA.2

Type:	single-seat multi-role combat aircraft
Powerplant:	one 9752kg (21,500lb) thrust Rolls-Royce Pegasus Mk 106 vectored thrust turbofan
Performance:	maximum speed 1185km/h (736mph) at sea level; service ceiling 15,545m (51,000ft); combat radius 185km (115 miles) on high-level CAP with 90-minute loiter on station
Weights:	empty 5942kg (13,100lb); maximum take-off 11,884kg (26,200lb) loaded
Dimensions:	wing span 7.70m (25ft 3in); length 14.17m (46ft 6in); height 3.71m (12ft 2in); wing area 18.68m² (201.1 sq ft)
Armament:	two 25mm (1in) Aden cannon; five external pylons with provision for AIM-9 Sidewinder, AIM-120 AMRAAM, and two Harpoon or Sea Eagle anti-ship missiles, up to a total of 3629kg (8000lb)

301
aerospatiale

France

At the end of the war in Europe, France's aircraft industry – which, in 1939, had begun to produce excellent combat aircraft – lay in ruins, its factories destroyed or dismantled, its designers scattered far and wide.

In seeking to establish a leading role in post-war aviation, therefore, France was faced with a mammoth two-fold task. The first priority was an industrial one, to rebuild the factories and reassemble the design bureaux; the second was of a purely technical nature, involving the production of new combat types to meet the demands of the French Air Force in the jet age. The second priority was much harder to achieve than the first. Although some French designers had made studies of jet aircraft projects in secret during the occupation, they lagged far behind the Germans and the Allies, from the viewpoint of both airframe and engine design, and at the war's end, even with the knowledge that turbojet-powered aircraft were the only answer to meeting future high-performance requirements, some designers persisted in launching new piston-engined projects that resulted only in a waste of time and a dissipation of resources.

Yet within a decade, thanks to the skill and ingenuity of French designers – notably Marcel Dassault, formerly Marcel Bloch, designer of the wartime Bloch MB.151 and 152 radial-engined fighters – coupled in the first instance with the proven reliability of licence-built British aero-engines, France was producing combat aircraft that were second to none.

Left: **Armed with the ASMP medium-range missile, the Mirage 2000N has taken over the nuclear deterrent role formerly assigned to the Mirage IVP. The Mirage 2000N equips EC 01.004 and 02.004 at Luxeuil and EC 03.004 at Istres.**

Dassault Mystère IVA

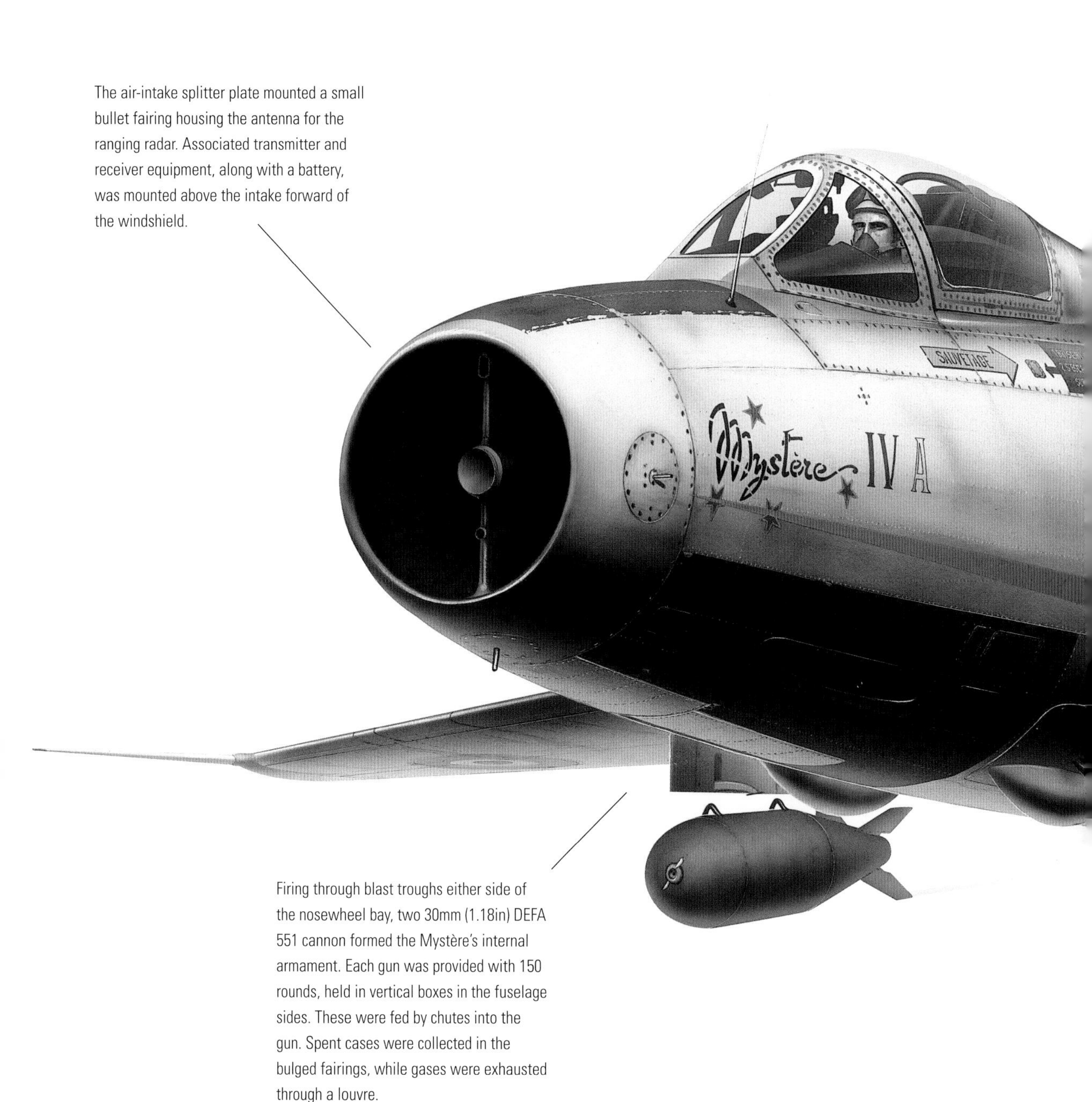

The air-intake splitter plate mounted a small bullet fairing housing the antenna for the ranging radar. Associated transmitter and receiver equipment, along with a battery, was mounted above the intake forward of the windshield.

Firing through blast troughs either side of the nosewheel bay, two 30mm (1.18in) DEFA 551 cannon formed the Mystère's internal armament. Each gun was provided with 150 rounds, held in vertical boxes in the fuselage sides. These were fed by chutes into the gun. Spent cases were collected in the bulged fairings, while gases were exhausted through a louvre.

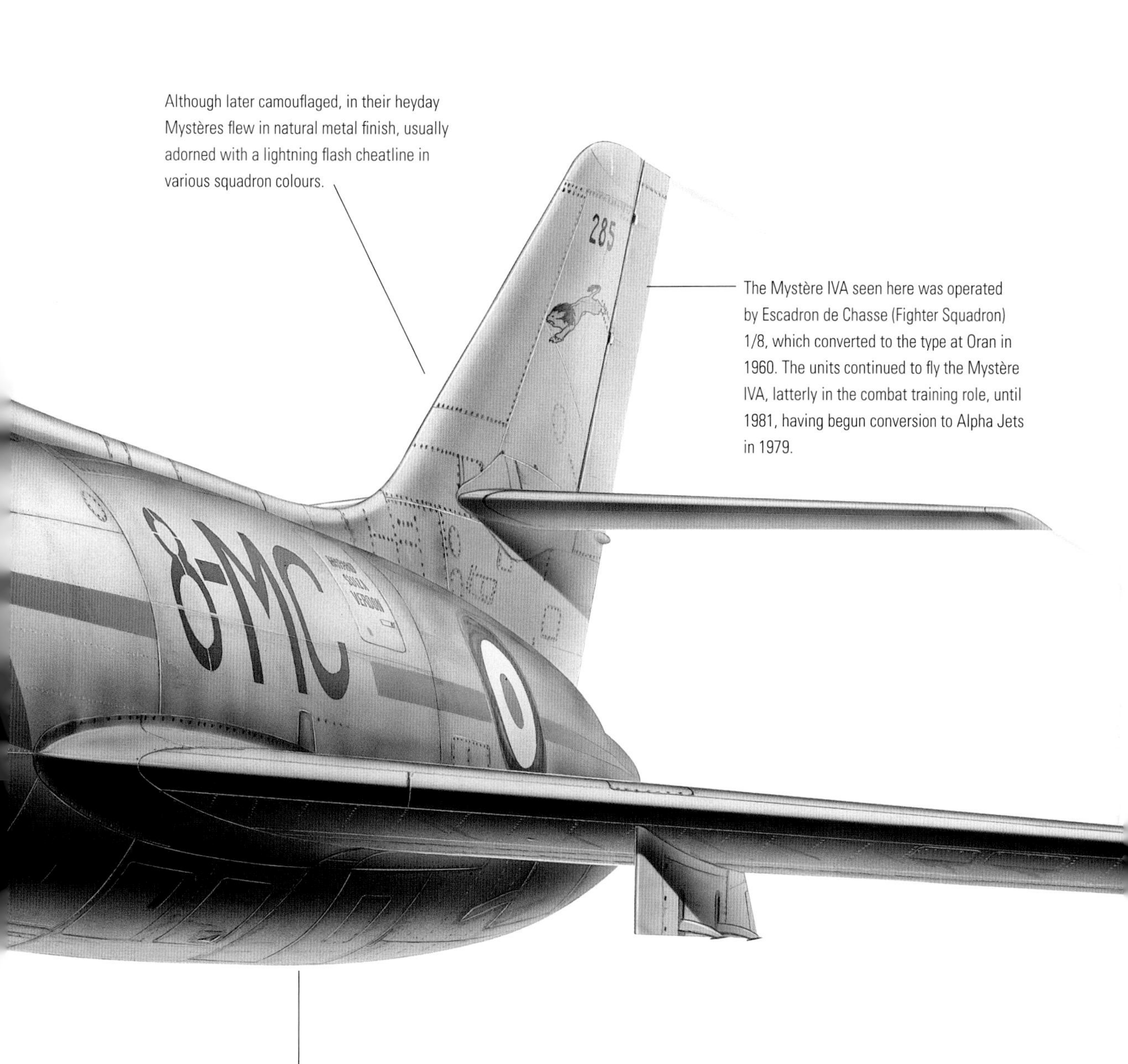

Although later camouflaged, in their heyday Mystères flew in natural metal finish, usually adorned with a lightning flash cheatline in various squadron colours.

The Mystère IVA seen here was operated by Escadron de Chasse (Fighter Squadron) 1/8, which converted to the type at Oran in 1960. The units continued to fly the Mystère IVA, latterly in the combat training role, until 1981, having begun conversion to Alpha Jets in 1979.

The rear fuselage of the Mystère was necessarily fat to house the large centrifugal-flow Tay/Verdon engine. Access to this was gained by removing the whole rear fuselage/tail assembly at a break line roughly level with the point where the wing trailing edge faired into the fuselage.

For the first few years of its post-World War II existence, the French Air Force had no alternative but to rely on foreign jet aircraft like the de Havilland Vampire for its first-line equipment, but on 28 February 1949 Avions Marcel Dassault flew the prototype of a straightforward, no-frills jet fighter, which it had begun as a private venture in November 1947. Powered by a Rolls-Royce Nene 102 turbojet, built under licence by Hispano-Suiza, the Dassault MD.450 Ouragan (Hurricane) became the first jet fighter of French design to be ordered in quantity, some 350 production aircraft being delivered to the French Air Force from 1952. The Ouragan was exported to India, where it was known as the Toofani (Whirlwind), and to Israel, which received 75 examples. Although the Ouragan was inferior to the MiG-15, the principal jet fighter type equipping the Egyptian Air Force at that time, it performed well in the ground attack role.

The Dassault MD.452 Mystère IIC, which flew for the first time on 23 February 1951, was a straightforward swept-wing version of the Ouragan. Some 150 Mystère IICs served with the French Air Force, and Israel had plans to purchase some in 1954/5, but in view of the type's poor service record – several of the earlier French machines having been lost through structural failure – it was decided to buy the much more promising Mystère IV instead.

The Dassault Mystère IV was unquestionably one of the finest combat aircraft of its era. Although developed from the Mystère IIC, it was in fact a completely new design. The prototype Mystère IVA flew for the first time on 28 September 1952, and early trials proved so promising that

Below: **The Dassault Mystère IV, seen here in prototype form, proved itself in combat while serving with the Israeli and Indian Air Forces. Although designed primarily as an interceptor, the type was readily adapted to the role of fighter-bomber.**

Above: **French Air Force Mystère IVAs were deployed to Israel during the Suez campaign of 1956, and were painted in Israeli markings. They were used in the air defence role, while French Air Force F-84F Thunderstreaks operated offensively against Egyptian airfields.**

the French Government placed an order for 325 production aircraft six months later in April 1953. The fighter was also delivered to India, and Israel acquired the first of 60 in April 1956, the type replacing the Gloster Meteor F.8 in Israeli Air Force service. Production of the Mystère IVA was completed in 1958, with the 421st aircraft. French Mystère IVA procurement totalled 241, of which 225 were paid for by the US under the offshore procurement scheme. Mystère IVAs were initially delivered to EC2, EC5 and EC12 for service in the air defence role, and subsequently to EC7 and EC8 in North Africa, where the aircraft was used primarily in the fighter-bomber role. In addition to these front-line Escadres de Chasse, the type was flown by training units Groupement Instruction 312 (École de l'Air) at Salon-de-Provence, which operated a flight until 1976, and Groupement École 314 (École de Chasse) at Tours (formerly based at Meknes, Morocco, until 1961), which retired its last aircraft in 1973.

Israel purchased 60 Mystère IVAs for the air defence role, the first 24 being delivered in May 1956 for service with No 101 Squadron in time for action in the Sinai (Suez) campaign, in which the type claimed seven kills. Another 36 arrived in August, including one fitted out for reconnaissance, allowing No 109 Squadron to re-equip in December. In the 1960s Mirage IIIs arrived, allowing the Mystères to be given a new role as fighter-bombers, seeing action in the Six-Day War of 1967. They were phased out in the 1970s. The Mystère IVA was also used by India, which followed up its order for the Toofani with the purchase of 110 Mystère IVAs, the first being issued to No 1 Squadron in 1957. The type saw extensive action in the 1965 war against Pakistan, used mainly in the attack role by Nos 3 and 31 Squadrons. The last aircraft was retired by No 31 Squadron in 1973.

A one-off variant of the Mystère IV, the Mystère IVB, was fitted with an afterburning Rolls-Royce RA7R turbojet and became the first French aircraft to exceed the speed of sound in level flight. It served as a test bed for the next Dassault fighter, the Super Mystère B.2. The Super Mystère was the Mystère IVA's transonic successor, with a thinner, more sharply swept wing, an improved air intake and modified cockpit. This type also saw service with the Israeli Air Force, destroying several Egyptian MiG-17s in the skirmishes that led up to the Six-Day War of June 1967 for the loss of six of their own number.

Specification: Dassault Mystère IVA	
Type:	single-seat fighter-bomber
Powerplant:	one 2850kg (6280lb) thrust Hispano-Suiza Tay/Verdon 250A turbojet
Performance:	maximum speed 1120km/h (696mph); service ceiling 13,750m (45,000ft); range 1320km (820 miles)
Weights:	empty 5875kg (12,950lb); maximum take-off 9500kg (20,950lb)
Dimensions:	wing span 11.10m (36ft 5in); length 12.90m (42ft 4in); height 4.40m (14ft 5in)
Armament:	two 30mm (1.18in) DEFA 551 cannon; up to 907kg (2000lb) of stores

Dassault Mirage III

The Mirage IIICZ was equipped with the Martin-Baker ZRM4 (Mk 4 series) ejection seat, which could not be used at speeds under 167km/h (104mph).

One of the reasons for the adoption of a delta wing for the Mirage III was the requirement for a supersonic thickness/chord ratio (4.5 per cent at the root and 3.5 per cent at the tips) without resorting to what were, at the time, complex manufacturing processes such as those associated with the Lockheed Starfire's ultra-thin wing.

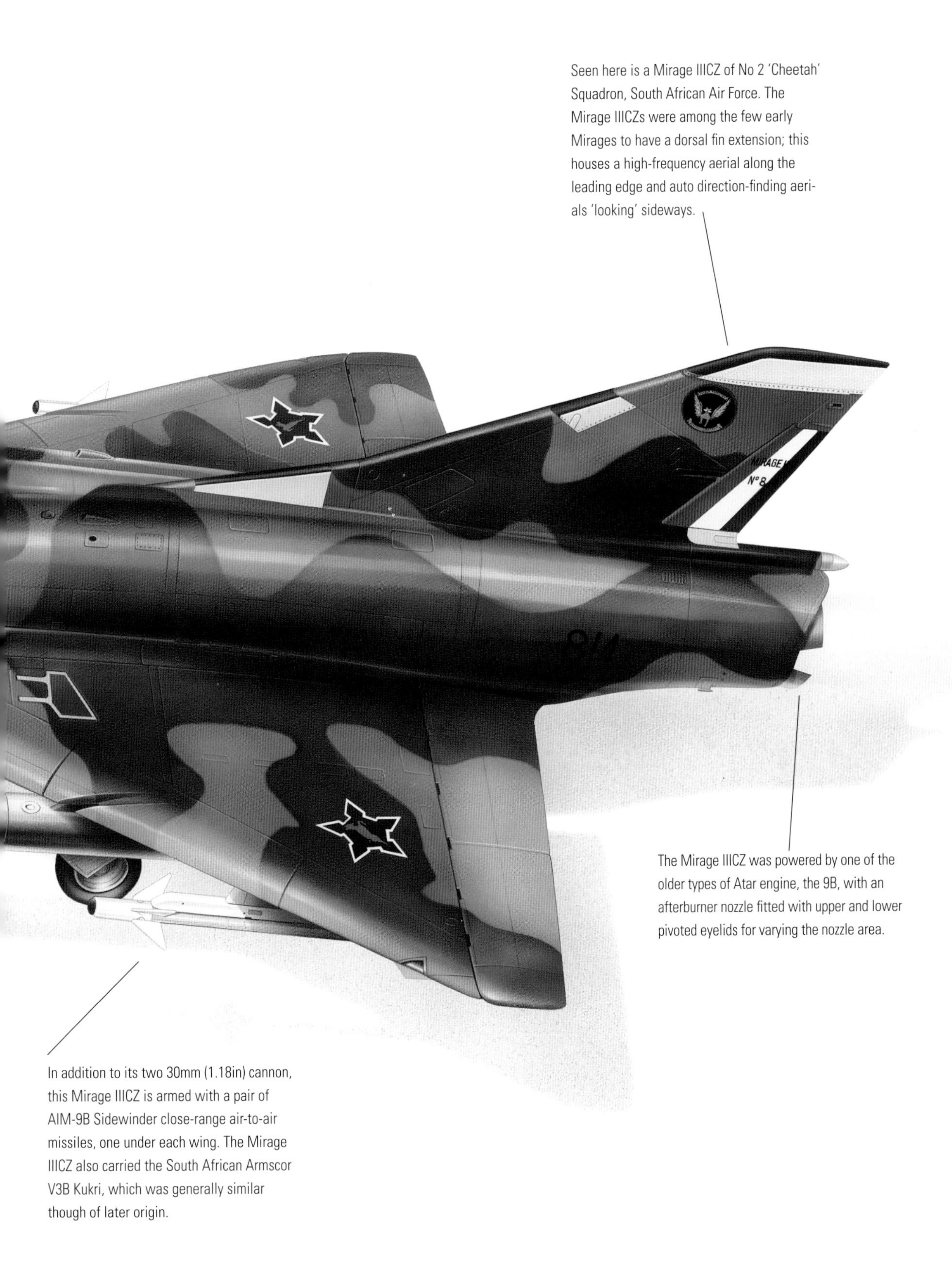

Seen here is a Mirage IIICZ of No 2 'Cheetah' Squadron, South African Air Force. The Mirage IIICZs were among the few early Mirages to have a dorsal fin extension; this houses a high-frequency aerial along the leading edge and auto direction-finding aerials 'looking' sideways.

The Mirage IIICZ was powered by one of the older types of Atar engine, the 9B, with an afterburner nozzle fitted with upper and lower pivoted eyelids for varying the nozzle area.

In addition to its two 30mm (1.18in) cannon, this Mirage IIICZ is armed with a pair of AIM-9B Sidewinder close-range air-to-air missiles, one under each wing. The Mirage IIICZ also carried the South African Armscor V3B Kukri, which was generally similar though of later origin.

One of the biggest success stories in post-1945 combat aircraft design, the Dassault Mirage III owed its origin to the Dassault MD550 Mirage I of 1954, which, together with the SE Durandal and the SO Trident, was a contender in a French Air Force competition for a lightweight high-altitude rocket-assisted interceptor capable of using grass strips. The MD550 Mirage I flew for the first time on 25 June 1955, powered by two Bristol Siddeley Viper turbojets. In May 1956 the aircraft reached Mach 1.15 in a shallow dive, and on 17 December 1956, with the additional boost of an SEPR 66 rocket motor, it attained Mach 1.3 in level flight. The Mirage I proved too small, however, to carry an effective war load and its twin Viper turbojets lacked the necessary power, so Dassault regarded it as a development aircraft for another proposal, the Mirage II, which was to be powered by two Turboméca Gazibo turbojets fitted with reheat, but then the company had a further change of mind. Before the Mirage I had even begun its transonic trials, Dassault decided to abandon the Mirage II project in favour of a larger version. The airframe was substantially redesigned and enlarged and fitted with a single afterburning SNECMA Atar 101G-2 turbojet. The new aircraft made its first flight on 17 November 1956 and exceeded Mach 1.5 in level flight on 30 January 1957.

Hopes that the new aircraft would be selected as an F-86 Sabre replacement for the Luftwaffe and other NATO air forces were shattered when the Lockheed F-104 Starfighter was chosen instead, and it became clear that the emphasis in the future would be on versatility. The French government accordingly instructed Dassault to proceed with a multi-mission version, the Mirage IIIA, the prototype of which

Below: **The Mirage IIIC shown here was deployed with EC 3/10 'Vexin' to Djibouti in the early 1980s and is illustrated as it appeared then, finished in what was officially described as a 'sand and chestnut' camouflage scheme.**

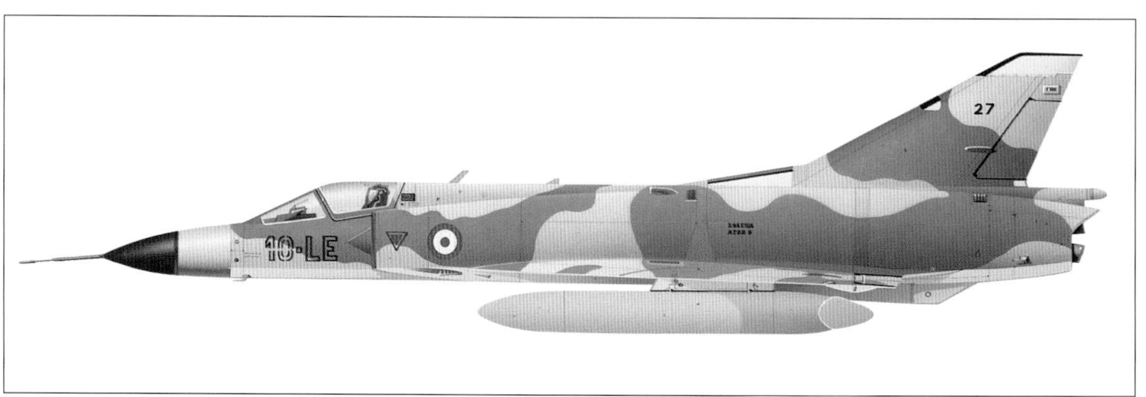

(Mirage IIIA-01) flew on 12 May 1958, and in a test flight on 24 October 1958 this aircraft exceeded Mach 2 in level flight at 12,500m (41,000ft). A pre-series batch of ten aircraft was built, powered by the SNECMA Atar 9B turbojet. The last six aircraft were equipped to production standard with the CSF Cyrano Ibis air-to-air radar. The Mirage IIIB was a two-seat version of the IIIA, with tandem seating under a one-piece canopy; the radar was deleted, but radio beacon equipment was fitted.

Although intended primarily as a trainer, the IIIB could also be configured for strike sorties, and carried the same air-to-air armament as its predecessor. The prototype flew for the first time on 20 October 1959,

Above: **A Mirage IIIE flying high over southern France. The IIIE variant first appeared in April 1961. This aircraft is carrying a Matra R.530 air-to-air missile on its centreline installation. This weapon was produced in both infrared and radar homing versions.**

followed by the first production model on 19 July 1962. The Mirage IIIC, which flew on 9 October 1960, was the first production version and was identical to the IIIA, with an Atar 9B3 turbojet and an SEPR 841 or 844 auxiliary rocket motor. One hundred Mirage IIICs were ordered by the Armée de l'Air, equipping the 2e and 13e Escadres de Chasse. 72 similar aircraft, without rocket motors or missiles, were supplied to the Israeli Air Force, first deliveries being made to No 101 Sqn in 1963. These aircraft were designated Mirage IIICJ and saw considerable action during the subsequent Arab–Israeli wars. Sixteen more aircraft of the IIIC series were supplied to South Africa as the Mirage IIICZ; South Africa took delivery of its first Mirage IIIs in December 1962, these aircraft entering service with No 2 'Cheetah' Squadron at Waterkloof in April 1963. Over the next decade these were supplemented by deliveries of Mirage IIIEZs, IIIDZs, IIID2Zs and IIIRZs. The SAAF also took delivery of three Mirage IIIBZ two-seaters, which carried the same armament as the IIIC. During the mid-1970s the SAAF's Mirage IIIs were gradually replaced by Mirage F.1s, and some of the later-model Mirage IIIs were rebuilt as the Atlas Cheetah.

The Mirage IIID was a two-seat version of the Mirage IIIO, manufactured under licence in Australia; the first of 16 ordered for the Mirage Operational Conversion Unit was assembled in Australia and delivered in November 1966. The Mirage IIIE was a long-range tactical strike variant, 453 examples being produced for the Armée de l'Air and

Below: **Mirage IIIEEs, supplied to Spain, were known locally as C.11s. Like many Mirage operators, the Spanish Air Force planned to upgrade its fleet, but budget cuts saw the type's premature retirement in 1992. It was replaced by the McDonnell Douglas F/A-18.**

further aircraft for export. The first of three prototypes flew on 5 April 1961, and the first production example was delivered in January 1964. The Mirage IIIE equipped eight squadrons of the Force Aérienne Tactique (FATAC), the French tactical air force, and was supplied to the air forces of Brazil, Lebanon, Argentina, South Africa, Pakistan, Libya, Spain and Switzerland. It was a version of the IIIE, the IIIO, that was manufactured under licence in Australia. The Royal Australian Air Force received 50 Mirage IIIO(F) interceptors, 50 Mirage IIIOA ground attack aircraft and the 16 Mirage IIID two-seat trainers mentioned above. All remaining Mirage IIIO(F)s were converted to IIIO(A) standard in 1976–80. The Australian Mirages served with No 3 Squadron at Butterworth, Malaysia, No 75 at Darwin and No 72 at Williamtown. In 1984 the RAAF began receiving F/A-18 Hornets to replace its Mirages, the last being retired in 1988.

The Mirage IIIEE (the second E denoting Espagne) came on to the Spanish Air Force's inventory in 1970 and equipped No 11 Wing (Nos 111 and 112 Squadrons at Manises), which operated 19 EEs and six two-seat EDs. The Mirage IIIP version for Pakistan saw action in the 1971 conflict with India. Another version of the Mirage IIIE was the IIIS, delivered to the Swiss Air Force. The Swiss Mirage IIIs remained in service for 35 years, being eventually replaced by the F-18 Hornet. The Mirage IIIR was the reconnaissance version of the IIIE, equipped with a battery of five OMERA Type 31 cameras in place of the nose radar. In Argentine service, Mirages saw combat against the British task force in the campaign to recapture the Falkland Islands in 1982.

In October 1965 the French government awarded Dassault a contract for the development of a variable-geometry version of the Mirage, the Mirage G. The development of this aircraft was to proceed in parallel with another project, the Anglo-French Variable Geometry Aircraft, which was later abandoned. The Mirage G flew for the first time at Istres on 18 November 1967, and within two months of this debut the aircraft was reaching a level speed of over Mach 2. This test aircraft was destroyed in an accident after completing some 400 flying hours. Two prototypes of a smaller, twin-engined VG aircraft, the Mirage G8, were subsequently ordered by the French Government; the first flew on 8 May 1971 and reached Mach 2.03 four days later, but the type was not adopted.

Specification: Dassault Mirage IIIE	
Type:	single-seat tactical strike aircraft
Powerplant:	one 6200kg (13,668lb) thrust SNECMA Atar 9C turbojet
Performance:	maximum speed 1390km/h (863mph) at sea level; service ceiling 17,000m (55,775ft); combat radius 1200km (745 miles) at low level with 907kg (2000lb) payload
Weights:	empty 7050kg (15,540lb); maximum take-off 13,500kg (29,760lb)
Dimensions:	wing span 8.22m (26ft 11in); length 16.50m (54ft 2in); height 4.50m (14ft 9in); wing area 35m² (376.7 sq ft)
Armament:	two 30mm (1.18in) DEFA cannon; provision for up to 3000kg (6612lb) of external stores, including special (i.e. nuclear) weapons

Dassault Mirage F.1

The Mirage F.1AZ (Z for Zuid Afrika) seen here served with No 1 Squadron, South African Air Force, at Hoedspruit AB. The SAAF's last Mirage F.1s were retired in 1997.

The F.1A fighter-bomber carried a small EMD Aida 2 ranging radar in the extreme nose. The radar had a fixed antenna and provided automatic search, acquisition, ranging and tracking for targets within its 16 degree field of view. Data was presented to the pilot in his gyro gunsight.

The bulge under the F.1AZ's nose housed a Thomson-CSF TMV360 laser rangefinder, which provided accurate distance measuring for the ground attack role.

The fin-mounted forward- and rearward-facing antennae for the Thomson-CSF BF radar warning receiver. Sideways cover was provided by disc antennae flush with the fin sides.

The F.1AZ's principal armament comprised two internal cannon, with most stores carried on multiple dispensers on the centreline. As shown here, the F.1AZ could be fitted with wingtip launch rails for the V3B Kukri or V3C Darter indigenous air-to-air missile.

Above: **The undernose bulge seen on this South African Air Force Mirage F.1AZ houses a Thomson-CSF TMV-360 laser rangefinder, which provides accurate distance measuring for the ground attack role. The aicraft also has a fixed refuelling probe.**

The Mirage F.1 single-seat strike fighter was developed as a private venture. Powered by a SNECMA Atar 9K, the prototype flew for the first time on 23 December 1966; it exceeded Mach 1 during its fourth flight on 7 January 1967, but was lost in a fatal accident on 18 May that year. In September 1967 three pre-series aircraft and a structural test airframe were ordered by the French Government, and the first of these completed the first phase of its flight test programme in June 1969. The first production aircraft entered service with the 30e Escadre at Reims early in 1974. Variants produced included the F.1A ground attack aircraft, the F.1C interceptor and the F.1B two-seat trainer. The Mirage F.1's wing, a departure from the traditional Dassault delta format, was fitted with elaborate high-lift devices which permitted the aircraft to take off and land within 500–800m (1600–2600ft) at average combat mission weight. Primary role of the Mirage F.1 was all-weather interception at any altitude, and the original production version used the same weapon systems as the Mirage III. The Mirage F.1 was the subject of large overseas export orders, notably to countries in the Middle East.

Specification: Dassault Mirage F.1AZ	
Type:	single-seat multi-role fighter/attack aircraft
Powerplant:	one 7200kg (15,876lb) thrust SNECMA Atar 9K-50 turbojet
Performance:	maximum speed 2350km/h (1460mph) at high altitude; service ceiling 20,000m (65,615ft); range 900km (560 miles) with maximum load
Weights:	empty 7400kg (16,317lb); maximum take-off 15,200kg (33,510lb)
Dimensions:	wing span 8.40m (27ft 6in); length 15.00m (49ft 2in); height 4.50m (14ft 9in); wing area 25m² (269.11 sq ft)
Armament:	two 30mm (1.18in) DEFA 553 cannon; up to 6300kg (13,891lb) of external stores

Export Mirage F.1s were distinguished by a suffix letter, for example the F.1CK for Kuwait. Ironically, the type was also supplied to Iraq as the Mirage F.1Q. Iraq's changing political allegiances during Saddam Hussein's regime were reflected in the widely varying types operated by the Iraqi Air Force, a fact well demonstrated during the war with Iran in the 1980s and the 1991 Gulf War. During the first conflict, France was a major supplier of Iraqi aircraft, delivering no fewer than 89 Mirage F.1s. Included in this total was a batch of 29 F.1EQ5s with Agave radar for over-water operations and armed with Exocet anti-shipping missiles. After much delay these aircraft were delivered in October 1984, and became operational in the following year.

The Mirage F.1AZ was a version for South Africa, which took delivery of both the attack-dedicated F.1AZ and the radar-equipped F.1CZ fighter. From an original order for 48 aircraft – including 32 F.1AZs – the first two aircraft were delivered on 5 April 1975. Surrounded by a cloak of secrecy, these aircraft (both F.1CZs) were delivered inside a South African Air Force Hercules. The remaining CZs were all delivered by July that year. During October 1975 F.1CZs appeared in an air display, but were not publicly acknowledged as being in SAAF service for a further 13 months. The F.1AZs were delivered between November 1975 and October 1976, with their operating unit, No 1 Squadron, leaving its Canadair CL-13 Sabres at Pietersburg to re-equip with the F.1AZ at Waterkloof. Again, secrecy covered the SAAF's F.1 acquisitions and No 1 Squadron was not able to show its aircraft publicly until February 1980. Both variants were involved in offensive actions in Angola and in anti-guerrilla strikes in the administered territory of South-West Africa before it gained independence as Namibia in 1989. In the course of these operations two Angolan MiG-21s were shot down by F.1CZs, while at least one CZ was damaged by an Angolan SAM. The SAAF began retiring its F.1CZs in September 1992, and the F.1AZs in November 1997.

Below: **Pictured in flight over the Alps, these Mirage F.1 CT aircraft of the French Air Force are upgrades of the basic design, featuring improved avionics and internal fuel tankage. Some aircraft were equipped with ground-data links for reconnaissance missions.**

Dassault Super Etendard

In service for more than 20 years, the Super Etendard has lasted far beyond its planned career and is to be replaced by the Dassault Rafale. The example seen here belonged to the now defunct Flotille 14F, which was based at Landivisiau.

Improvements in performance over the Etendard were made possible by the extra 336kg (741lb) of thrust obtained by using the 8K-50 version of SNECMA's Atar turbojet. It was basically the same engine as that installed in the Mirage F.1, but with an augmented jet pipe. It was less fuel-efficient than its predecessor, however, compelling the Super Etendard to carry extra fuel tanks as standard.

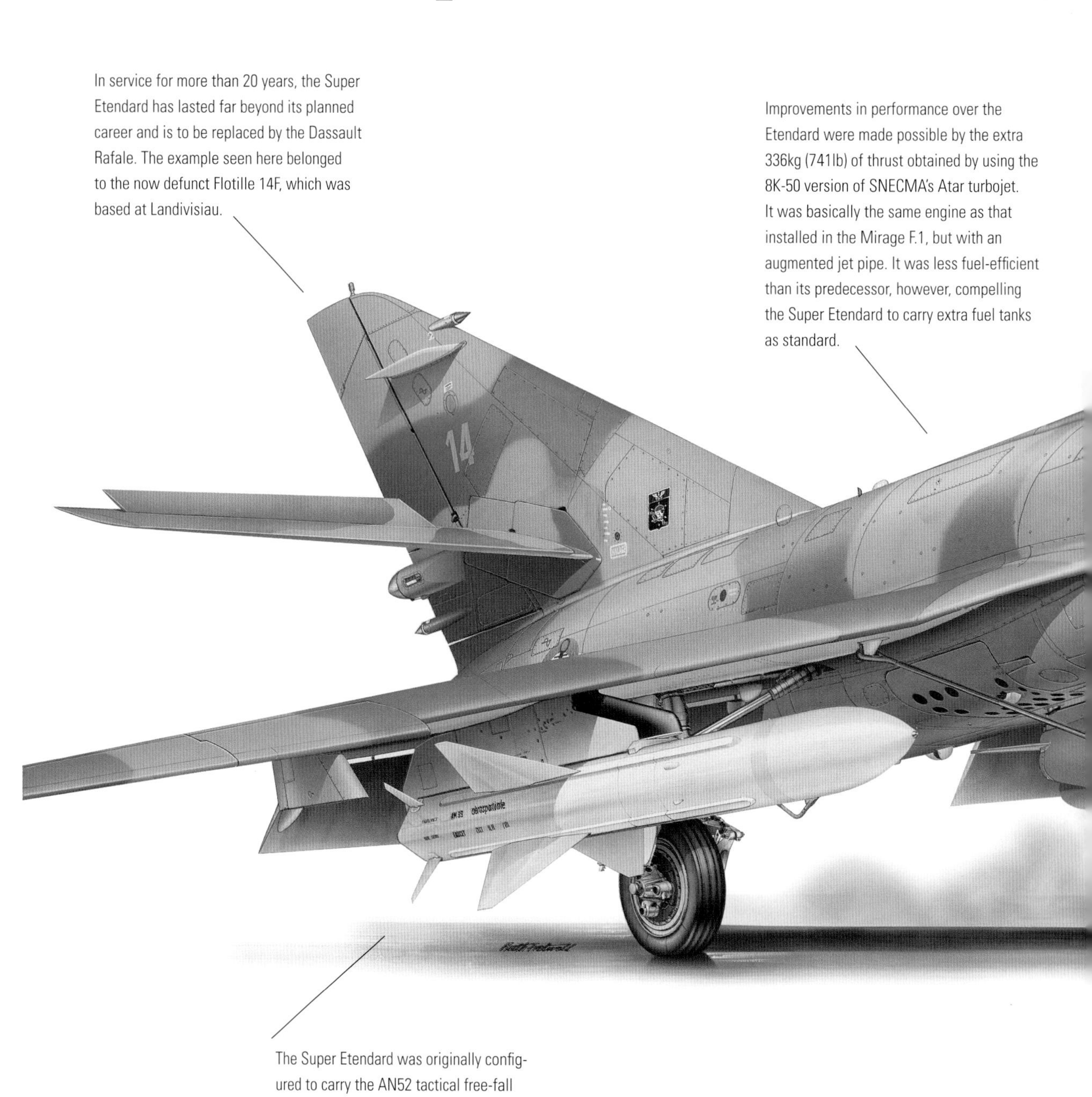

The Super Etendard was originally configured to carry the AN52 tactical free-fall nuclear weapon, which had a yield of 15kT and which was replaced in service by the medium-range ASMP (Air–Sol Moyen Portée) nuclear missile. This aircraft is carrying an Exocet AM39.

The Super Etendard's Thomson-CSF/ESD Agave radar is a simple lightweight set, able to detect an object the size of a patrol boat at about 40km (25 miles) and a fighter at 19km (12 miles). It is controlled by a sidestick on the left side of the cockpit.

Originally designed to meet the requirements of a mid-1950s tactical strike fighter contest – which it lost to the Fiat G.91 – the Dassault Etendard (Standard) showed such outstanding qualities that a development contract was awarded on behalf of the French Navy, which at that time was looking for a strike aircraft capable also of high-altitude interception. The navalized prototype Etendard IVM-01 flew on 21 May 1958, powered by a SNECMA Atar 8B turbojet, and began service trials in the following October. The first of 69 production Etendard IVMs was delivered to the naval air wing, the Aéronavale, on 18 January 1962, being followed into service by the Etendard IVP, an unarmed reconnaissance/tanker variant. It had originally been intended to replace the Etendard with a naval version of the SEPECAT Jaguar, but this was rejected by the Aéronavale and an advanced version of the Etendard was accepted instead. The Dassault Super Etendard, which first flew on 28 October 1974, was fitted with a SNECMA Atar 8K-50 turbojet and was intended for the low-level attack role, primarily against shipping.

Fourteen Super Etendards were supplied to Argentina from 1981, and the five that had been delivered by the time of the Falklands War of May–June 1982, armed with the Exocet ASM, proved highly effective against British vessels, sinking the Type 42 destroyer HMS *Sheffield* and the container ship *Atlantic Conveyor*. Having achieved a 'combat proven' status thanks to these two devastating attacks, the Super Etendard went on to see further action in Iraqi service during the Iran–Iraq 'tanker war' of the 1980s.

The Super Etendard's primary anti-shipping weapon was the Aérospatiale Exocet AM39, one of which was usually carried under the starboard inner wing pylon, counterbalanced by a fuel tank to port. The first ship-launched Exocet MM38s entered service in 1975, followed

Below: **In addition to a wide range of conventional weapons, the Super Etendard could be configured to carry the AN52 15kT tactical nuclear weapon. Built-in armament comprised two 30mm (1.18in) DEFA cannon, with MATRA Magic AAMs carried for air defence.**

by the AM39 air-to-surface variant in 1979, initially for carriage on the Super Etendard and later the Mirage F.1. Five were launched by the Argentine Navy's Super Etendards of the 2 Escuadrille de Caza y Ataque; fortunately for the British, these represented the Navy's entire stockpile of this formidable weapon. At least 100 were launched by Iraqi Super Etendards and Mirage F.1EQs during the 1980–88 war with Iran, two missiles launched by a Mirage being responsible for an accidental attack on the frigate USS *Stark* in May 1987. French Super Etendards have also been involved in combat operations over the former Yugoslavia and in 'police' actions in such trouble spots as Lebanon.

French Navy Super Etendards serving on the carriers *Charles de Gaulle, Foch* and *Clemenceau* are scheduled to be replaced by the Dassault Rafale by 2010. Meanwhile, the Super Etendard – popularly known as the 'Sue' – remains central to the French Navy's carrier-borne capability. Following re-equipment of Flottille 12F, a former F-8E (FN) Crusader unit, the Rafale-M will then replace the Super Etendard in Flottille 11F in 2006, with Flottille 17F completing the transition four years later. Another Super Etendard unit, Flottille 14F, was disbanded in 1991 as an economy measure.

Above: **A Super Etendard of the Argentine Navy. A superb attack aircraft, Argentine Super Etendards armed with Exocet sea-skimming missiles inflicted severe damage on the British task force during the Falklands war.**

Specification: Dassault Super Etendard	
Type:	single-seat carrier-borne strike and interceptor aircraft
Powerplant:	one 5000kg (11,025lb) thrust SNECMA Atar 8K-50 turbojet
Performance:	maximum speed 1180km/h (733mph) at low level; service ceiling 13,700m (44,950ft); combat radius 850km (528 miles) hi-lo-hi with one Exocet and two external tanks
Weights:	empty 6500kg (14,330lb); maximum take-off 12,000kg (26,460lb)
Dimensions:	wing span 9.60m (31ft 6in); length 14.31m (46ft 11in); height 3.86m (12ft 8in); wing area 28.4m² (305.7 sq ft)
Armament:	two 30mm (1.18in) DEFA cannon; provision for up to 2100kg (4630lb) of external stores; two Exocet ASMs; Matra Magic AAMs

Dassault Mirage 2000C

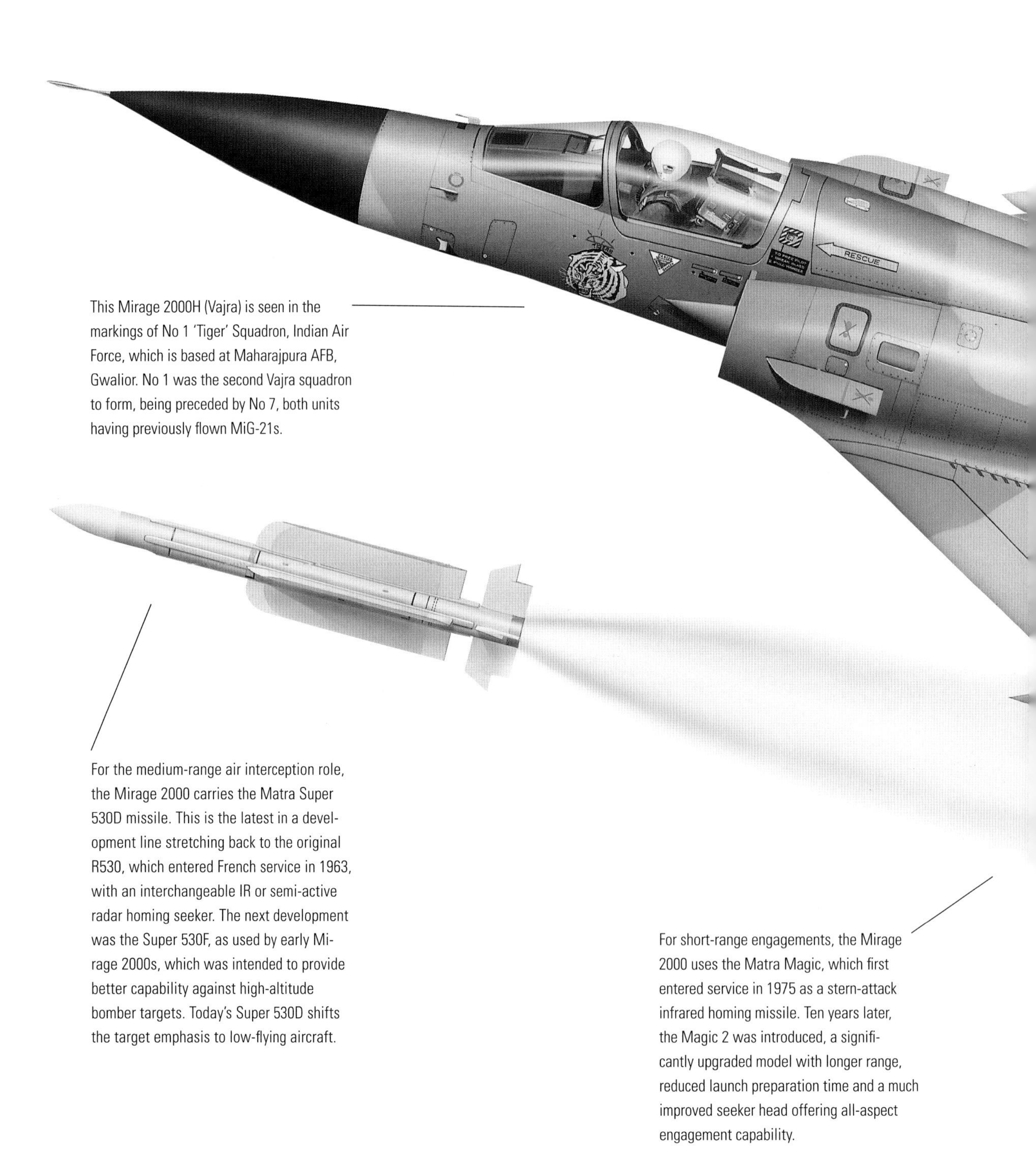

This Mirage 2000H (Vajra) is seen in the markings of No 1 'Tiger' Squadron, Indian Air Force, which is based at Maharajpura AFB, Gwalior. No 1 was the second Vajra squadron to form, being preceded by No 7, both units having previously flown MiG-21s.

For the medium-range air interception role, the Mirage 2000 carries the Matra Super 530D missile. This is the latest in a development line stretching back to the original R530, which entered French service in 1963, with an interchangeable IR or semi-active radar homing seeker. The next development was the Super 530F, as used by early Mirage 2000s, which was intended to provide better capability against high-altitude bomber targets. Today's Super 530D shifts the target emphasis to low-flying aircraft.

For short-range engagements, the Mirage 2000 uses the Matra Magic, which first entered service in 1975 as a stern-attack infrared homing missile. Ten years later, the Magic 2 was introduced, a significantly upgraded model with longer range, reduced launch preparation time and a much improved seeker head offering all-aspect engagement capability.

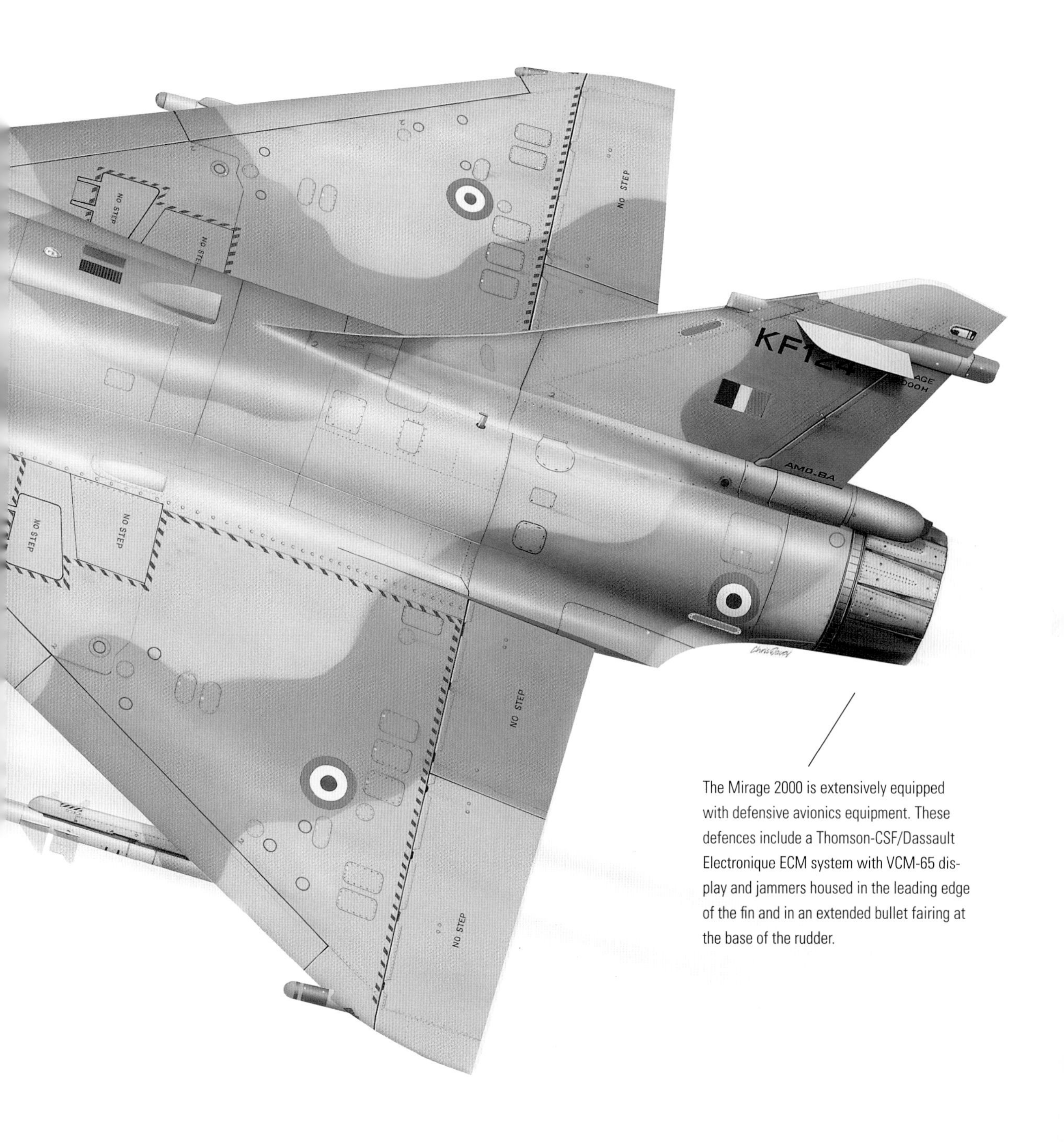

The Mirage 2000 is extensively equipped with defensive avionics equipment. These defences include a Thomson-CSF/Dassault Electronique ECM system with VCM-65 display and jammers housed in the leading edge of the fin and in an extended bullet fairing at the base of the rudder.

The Mirage 2000, the first of the Mirage family to take advantage of 'fly-by-wire' technology, was designed as an interceptor to replace the Mirage F.1. Its development was started after the failure of a collaborative programme, the Anglo-French Variable Geometry Aircraft, and of a number of subsequent projects involving VG aircraft of purely French design. The last of these was the ACF (Avion de Combat Futur/Future Combat Aircraft), cancelled by the French Government in 1975. As a consequence, the Mirage 2000 was formally adopted by the French Government on 18 December 1975 as the primary French Air Force combat aircraft from the mid-1980s. Under Government contract, the aircraft was developed initially as an interceptor and air superiority fighter, powered by a single SNECMA M53 turbofan engine and with Thomson-CSF multi-mode Doppler radar. It was soon appreciated, however, that the design had sound possibilities for reconnaissance, close support, and low-altitude attack missions in areas to the rear of a battlefield, and so the emphasis switched to multi-role. Five prototypes were built, of which four single-seat multi-role models were funded by the French Government and one two-seater by the manufacturers. The single-seat model flew for the first time at Istres on 10 March 1978, only 27 months after the programme was launched in December 1975. The second flew on 18 September 1978, the third on 26 April 1979 and the fourth on 12 May 1980.

The two-seat version, the Mirage 2000B (the fifth prototype) flew on 11 October 1980, and like its single-seat counterparts achieved supersonic speed (between Mach 1.3 and 1.5) on its first flight. On the basis of structural testing, the Mirage 2000 airframe was approved for a load factor of +9g and a rate of roll of 270 degrees per second in both subsonic and supersonic flight, clean or with four air-to-air missiles. A SNECMA M53-2 engine was fitted for early prototype testing, and was replaced in 1980 by the

Below: **Because of the complexity of the third generation Mirage 2000, the French Air Force developed a fully combat-capable two-seat trainer concurrently with the single-seat 2000C. The result was the Mirage 2000B, seen here.**

M53-5, which also powered initial production aircraft. The first prototype was subsequently re-engined with the more powerful M53-P2, as intended for later production aircraft, and in this revised form it made its first flight on 1 July 1983.

The initial production Mirage 2000C-1 made its first flight on schedule on 20 November 1982, and the first production

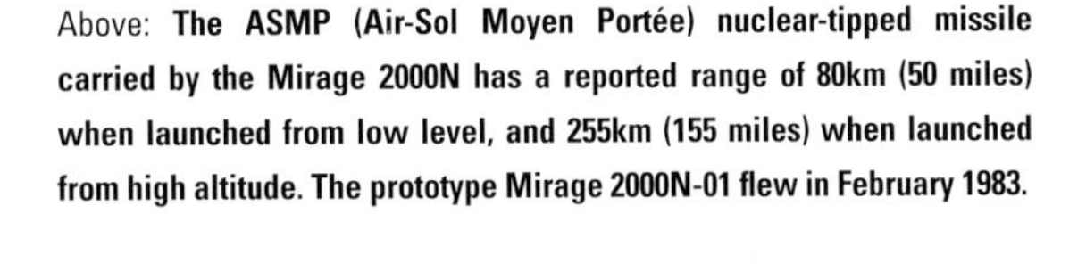
Above: **The ASMP (Air-Sol Moyen Portée) nuclear-tipped missile carried by the Mirage 2000N has a reported range of 80km (50 miles) when launched from low level, and 255km (155 miles) when launched from high altitude. The prototype Mirage 2000N-01 flew in February 1983.**

two-seat Mirage 2000B flew on 7 October 1983. The first unit to become operational with the Mirage 2000C-1 was Escadre de Chasse 1/2 'Cigognes' at Dijon on 2 July 1984. The Mirage 2000N, first flown on 2 February 1983, was developed as a replacement for the Mirage IIIE and is armed with the ASMP medium-range nuclear missile. This version is strengthened for operations at 1110km/h (690mph) at 60m (200ft). Seventy-five production aircraft were delivered from 1987. Like its predecessors, the Mirage 2000 has been the subject of substantial export orders from Abu Dhabi, Egypt, Greece, India and Peru. In Indian Air Force service the aircraft, designated Mirage 2000H, is known as the Vajra (Thunderstreak).

Specification: Dassault Mirage 2000C	
Type:	single-seat air superiority and attack fighter
Powerplant:	one 9700kg (21,388lb) thrust SNECMA M53-P2 turbofan
Performance:	maximum speed 2338km/h (1452mph) at high altitude; service ceiling 18,000m (59,055ft); range 1480km (920 miles) with 1000kg (2205lb) payload
Weights:	empty 7500kg (16,537lb); maximum take-off 17,000kg (37,485lb)
Dimensions:	wing span 9.13m (29ft 11in); length 14.36m (47ft 1in); height 5.20m (17ft 1in); wing area 41.0m² (441.3 sq ft)
Armament:	two DEFA 554 cannon; provision for up to 6300kg (13,890lb) of external stores

Dassault Rafale

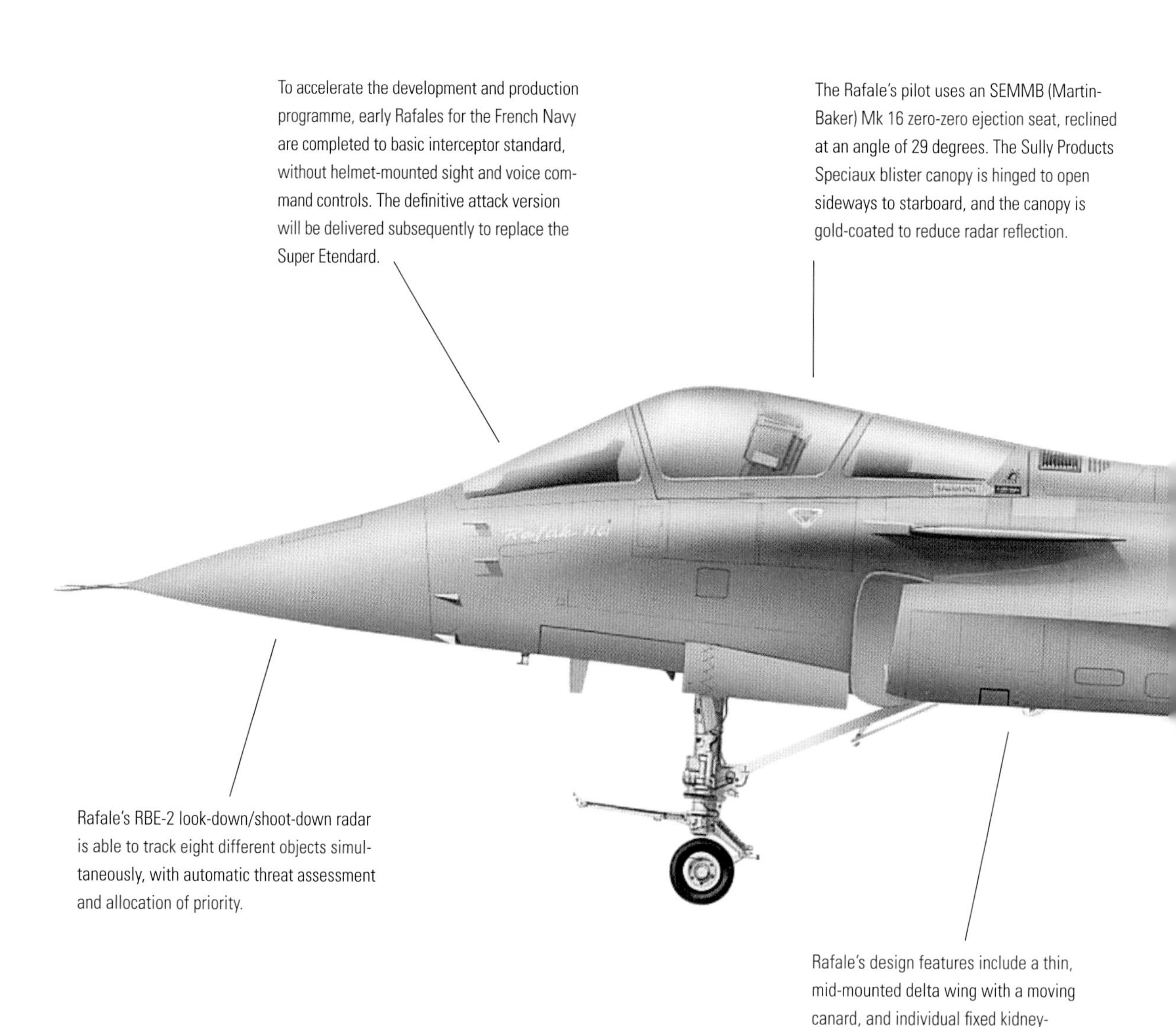

The Navy's Rafale-M has 80 per cent structural and equipment commonality with the French Air Force's Rafale-C, as well as 95 per cent systems commonality. The Navy's financial share of the development programme was cut from 25 to 20 per cent in 1991.

The hydraulically retracting tricycle-type landing gear is supplied by Messier Dowty, with single main wheels and twin, hydraulically steerable nosewheels. The undercarriage retracts forward and is designed to withstand a vertical impact speed of 3.0m (10ft) per second, or 6.5m (21ft) per second in the naval version, without flare-out.

France, originally a member of the European consortium that was set up to develop Eurofighter, decided to withdraw at an early stage and develop her own agile combat aircraft for the 21st century. The result was the Dassault Rafale (Squall). Known originally as the ACX (Avion de Combat Experimental), the main characteristics of the prototype were revealed in 1983, at which time it was announced that the type would replace the SEPECAT Jaguar in French Air Force service sometime in the 1990s, and that as the ACM (Avion de Combat Marine) it would form a major component of the air groups to be formed for deployment on the French Navy's new generation of nuclear-powered aircraft carriers. On the basis of an airframe with overall dimensions little greater than those of the Mirage 2000, Dassault set out to produce a multi-role aircraft capable of destroying anything from supersonic aircraft to helicopters in the air-to-air role, and able to deliver at least 3500kg (7715lb) of bombs or modern weapons on targets up to 650km (400 miles) from its base. The ability to carry at least six air-to-air missiles, and to fire them in rapid succession, was considered essential, together with the ability to launch electro-optically guided and advanced 'fire and forget' stand-off air-to-surface weapons. High manoeuvrability, high angle of attack flying capability under combat conditions, and optimum low-speed performance for short take-off and landing were basic design aims. This led to a choice of a compound-sweep delta wing, a large active canard foreplane mounted higher than the mainplane, twin engines, air intakes of new design in a semi-ventral position, and a single fin. To ensure a thrust-to-weight ratio far superior to one, it was decided to make extensive use throughout the

Below: **Dassault flew the first production Rafale, B.301, on 24 November 1998. The Rafale was designed to replace the French Air Force's fleet of SEPECAT Jaguars. It was originally expected that up to 250 aircraft would be ordered, but this figure was cut drastically.**

airframe of such composites as carbon, Kevlar and boron fibres, and of aluminium-lithium alloys, as well as the latest manufacturing techniques such as superplastic forming/diffusion bonding of titanium components. Ergonomic cockpit studies suggested that the pilot's seat should be reclined at an angle of 30 to 40 degrees during flight testing, and that equipment should include a sidestick controller, a wide-angle holographic head-up display, an eye-level display collimated to infinity (eliminating the need to refocus from the HUD to the instrument panel) and lateral multi-function colour displays.

A full-scale mock-up of the original ACX design was exhibited at the 1983 Paris Air Show, but the new model, displayed at the same venue two years later, revealed a number of significant refinements. In particular, Dassault-Breguet had been able to achieve improved flow into the engine air intakes, and greater efficiency at high angles of attack, by modifying the lower fuselage cross-section to a V-shape, enabling it to dispense with centrebodies and

Above: **The Rafale-A, pictured here, was built as a technology demonstrator, in the same way that the British Aerospace EAP was built as the demonstrator for the Eurofighter programme. The aircraft first flew on 4 July 1986, three years after the project was launched.**

other moving parts. The size of the fin was also greatly reduced. The definitive design that now emerged revealed a cantilever mid-wing monoplane of compound delta planform, with most of the wing components made from carbon fibre, including three-segment full-span elevons on each trailing edge.

As with Eurofighter, a technology demonstrator was built; known as Rafale-A, this flew for the first time on 4 July 1986. Powered by two SNECMA M88-2 augmented turbofans, each rated at 7450kg (16,425lb) thrust with reheat, Rafale is produced in three versions, the Rafale-C single-seat multi-role aircraft for the French Air Force, the two-seat Rafale-B, and the navalized Rafale-M. It incorporates digital fly-by-wire, relaxed stability and

electronic cockpit with voice command. The fly-by-wire system embodies automatic self-protection functions to prevent the aircraft from exceeding its limits at all times. Functional reconfiguration of the system in case of failure is also embodied, and provision has been made for the introduction of fibre optics to enhance nuclear hardening. Wide use of composites and aluminium-lithium has resulted in a seven to eight per cent weight saving. In the strike role, Rafale can carry one Aérospatiale ASMP stand-off nuclear bomb; in the interception role, armament is up to eight AAMs with either IR or active homing; and in the air-to-ground role a typical load is sixteen 227kg (500lb) bombs, two AAMs and two external fuel tanks. The aircraft is compatible with the full NATO arsenal of air–air and air–ground weaponry. Built-in armament comprises one 30mm (1.18in) DEFA cannon in the side of the starboard engine duct. The aircraft has a maximum level speed of Mach 2 at altitude and 1390km/h (863mph) at low level. France, which plans to have 140 Rafales in service by 2015 (in 2002 orders stood at 60 aircraft for the Air Force, and 24 of the Rafale-M variant for the Navy) sees the aircraft as vital to the defence of her territory. Low-level penetration combat radius with twelve 250kg (550lb) bombs, four AAMs and three external fuel tanks is 1055km (570 nautical miles). Dassault has launched a major export drive for Rafale, with Singapore, Saudi Arabia, South Korea and the United Arab Emirates all potential customers, but Eurofighter and Sweden's Gripen are powerful rivals.

Left: **The Rafale-M is the naval version of the Rafale, and was intended to replace the ageing F-8E Crusader in service with the French Navy. As with the French Air Force order, the number of naval Rafales scheduled to enter service will be significantly reduced.**

Specification: Dassault Rafale-C	
Type:	single-seat multi-role combat aircraft
Powerplant:	two 7450kg (16,427lb) SNECMA M88-2 turbofans
Performance:	maximum speed 2130km/h (1323mph) at high level; service ceiling classified; combat radius 1854km (1152 miles) air–air mission
Weights:	empty 9800kg (21,609lb); maximum take-off 19,500kg (42,990lb)
Dimensions:	wing span 10.90m (35ft 9in); length 15.30m (50ft 2in); height 5.34m (17ft 6in)
Armament:	one 30mm (1.18in) DEFA 791B cannon; up to 6000kg (13,230lb) of external stores

International

International cooperation in the design and development of military jet aircraft began within the framework of European NATO.

In the 1950s there was a wide diversification of aircraft types within the European air forces; American and British equipment predominated, although France was still a full NATO partner and some excellent designs were emerging from the Italian aircraft industry. The diversification of types and equipment, however, caused not a few operating problems: not only were the aircraft and their equipment different, but in many cases so was the ordnance they carried. NATO planners made strenuous efforts to achieve commonality, but it was a long and difficult path, made more tortuous by the determination of the French and British to pursue their own defence policies.

It was the cancellation of the BAC TSR-2 in 1965 that provided the spur to collaboration in Europe, first with the Anglo-French Variable Geometry Aircraft project, soon abandoned, and then with the MRCA, developed by Britain, Germany and Italy into the splendid Tornado. A couple of decades on the same consortium of nations, plus Spain, is fielding the Eurofighter Typhoon, after a development programme made protracted not by technical problems, but by politics – the factor that, more than any other, has bedevilled combat aircraft development since World War II. In the 1950s, for example, muddled political thinking, rather than concerns over lack of funding, led to the cancellation of several promising military aircraft projects in Britain, the misconception being that the day of the manned combat aircraft was approaching its end.

Left: **Head-on view of Eurofighter Typhoon DA.1 taxiing out to Manchings runway with test pilot Keith Hartley at the controls. The road that led to Eurofighter began with the cancellation of Britain's TSR-2 in 1965, spurring international co-operation.**

SEPECAT Jaguar

The characteristic wedge or 'chisel' shape of the Jaguar's nose encloses the aircraft's Ferranti laser rangefinder and marked target seeker.

As the Jaguar is a strike aircraft, its single-seat cockpit does not require the all-round visibility of a dedicated air superiority fighter. Internally, the cockpit is old-fashioned, having been designed long before the era of digital displays.

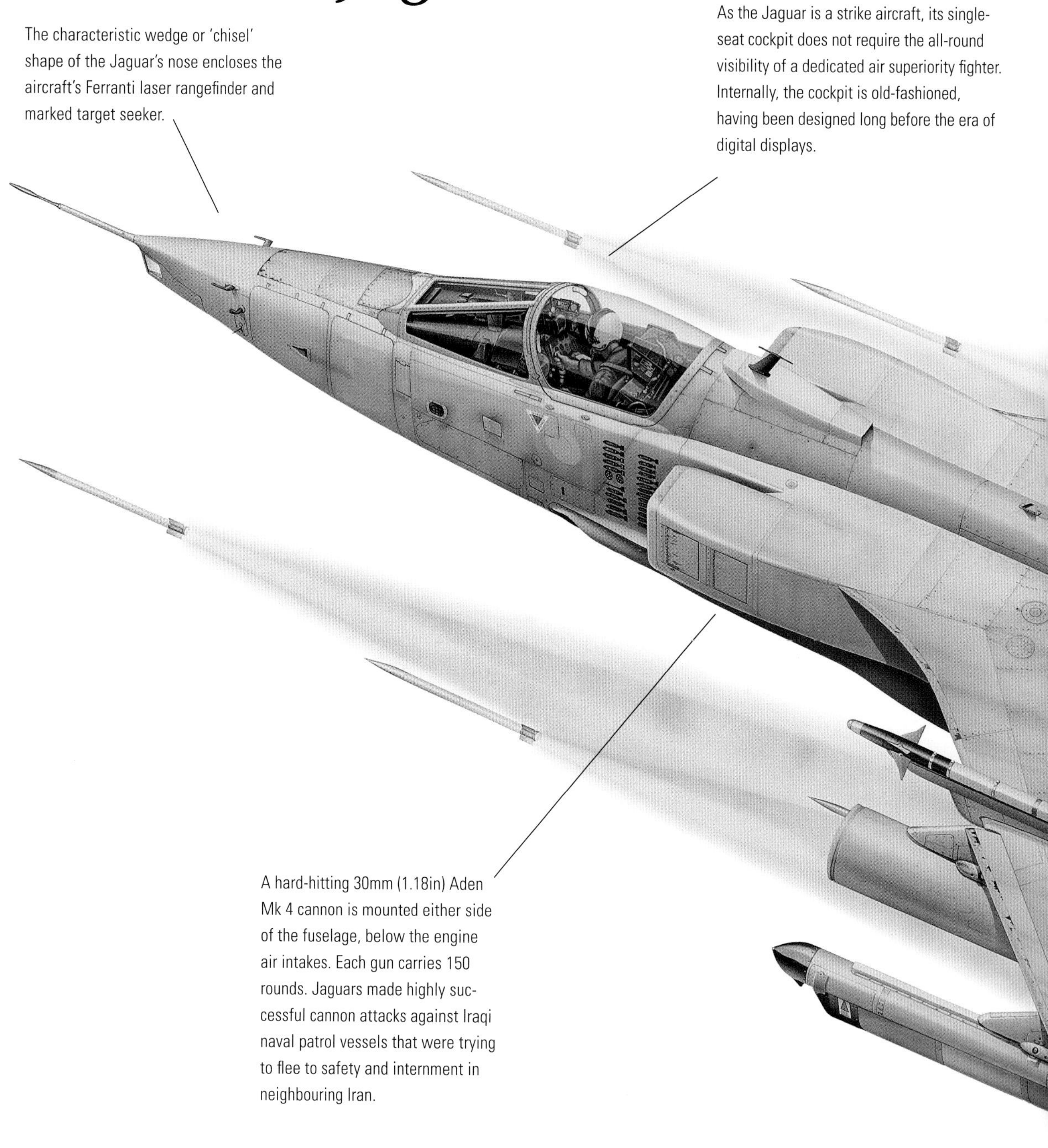

A hard-hitting 30mm (1.18in) Aden Mk 4 cannon is mounted either side of the fuselage, below the engine air intakes. Each gun carries 150 rounds. Jaguars made highly successful cannon attacks against Iraqi naval patrol vessels that were trying to flee to safety and internment in neighbouring Iran.

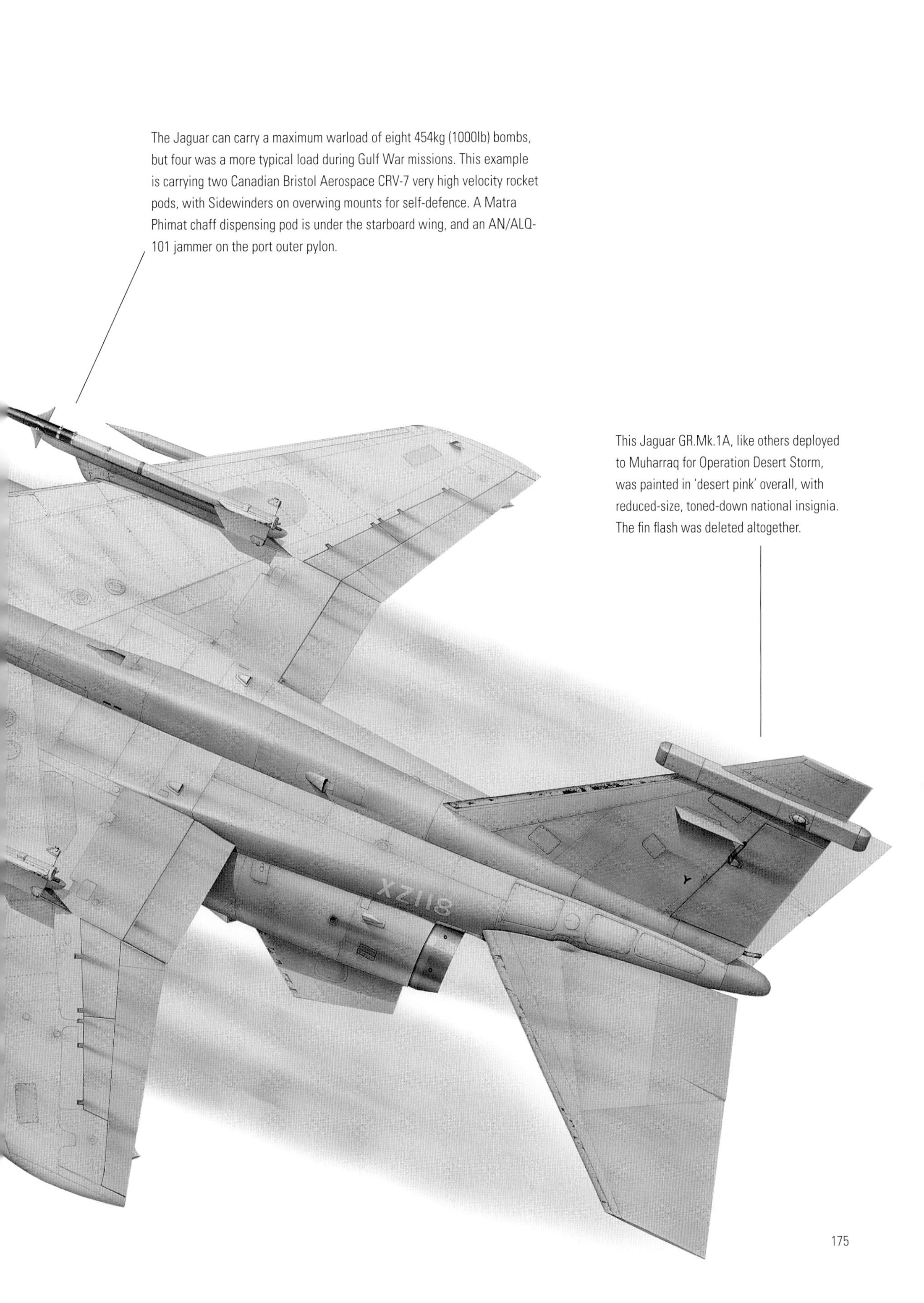

The Jaguar can carry a maximum warload of eight 454kg (1000lb) bombs, but four was a more typical load during Gulf War missions. This example is carrying two Canadian Bristol Aerospace CRV-7 very high velocity rocket pods, with Sidewinders on overwing mounts for self-defence. A Matra Phimat chaff dispensing pod is under the starboard wing, and an AN/ALQ-101 jammer on the port outer pylon.

This Jaguar GR.Mk.1A, like others deployed to Muharraq for Operation Desert Storm, was painted in 'desert pink' overall, with reduced-size, toned-down national insignia. The fin flash was deleted altogether.

Above: **A Jaguar GR.1 from No 6 Squadron, RAF Coltishall, over the flat landscape of East Anglia. The aircraft is carrying bombs, chaff and flare dispensers, and a Westinghouse ECM pod on the outer port wing station. The Jaguar gave excellent service in the 1991 Gulf War.**

Developed jointly by the British Aircraft Corporation and Breguet (later Dassault-Breguet) under the banner of SEPECAT (Société Européenne de Production de l'Avion École de Combat et Appui Tactique), the Jaguar emerged from protracted development as a much more powerful and effective aircraft than originally envisaged. The first French version to fly, in September 1968, was the two-seat E model, 40 being ordered by the French Air Force, followed in March 1969 by the single-seat Jaguar A tactical support aircraft. Service deliveries of the E began in May 1972, the first of 160 Jaguar As following in 1973. The British versions, known as the Jaguar S (strike) and Jaguar B (trainer), flew on 12 October 1969 and 30 August 1971 respectively, being delivered to the RAF as the Jaguar GR.Mk.1 (165 examples) and T.Mk.2 (38 examples). French Air Force Jaguars were fitted with a stand-off bomb release system consisting of two parts, the first of which was the ATLIS (Automatic Tracking Laser Illumination System) fire control equipment, contained in a pod mounted on the aircraft's centreline.

Above: **This Indian Air Force Jaguar is unusual in that it carries a maritime camouflage scheme in two shades of grey-blue. National insignia is applied in the form of saffron, white and green roundels and fin flashes, with RAF-style squadron markings and serial number.**

The pod contained a laser designator and a wide-angle TV camera with its field of view centred down the line of the laser beam. The assembly was stabilized and held steady regardless of aircraft movement. The second component of the system comprised a modular laser guidance unit called ARIEL, which was implanted in the nose-cones of rockets, missiles or bombs.

The British Jaguars were fitted with two weapon guidance systems: a Laser Ranging and Marked Target Seeker (LRMTS) and a Navigation and Weapon Aiming Subsystem (NAVWASS), both developed by Ferranti (later Marconi-Elliott). At the time of its delivery, the system, with its E3R inertial platform and MSC920M computer, seemed remarkable, offering a single-seat fighter pilot the best possible chance of making a first-pass attack without reference to tactical air navigation (TACAN) equipment or any other external aid, which might be unavailable in wartime. Steering commands were generated in the HUD, and aircraft position (which could be updated by observations in flight) was displayed on a moving map. The equipment marked a new level of navigational accuracy for non radar-equipped single-crew aircraft. During the mid-1980s NAVWASS was replaced in RAF and Omani Jaguars by the even better Ferranti FIN 1064, which also acted as a weapons delivery computer. Five black boxes were replaced by one, saving 50kg (110lb) in weight, quadrupling memory and saving much space. The pilot aligned his map on the TABS (Total Avionics Briefing System) digitizing map table in the briefing room, using a cursor to plot waypoints and enter them into a tiny portable data store, which could be plugged into the cockpit. TABS also supplied a hard copy on paper.

The Jaguar International, first flown in August 1976, was a version developed for the export market. It was purchased by Ecuador (12), Nigeria (18) and Oman (24) and was licence-built in India by HAL (98, including 40 delivered by BAe). A French plan to produce a carrier-borne version of the Jaguar was abandoned, the Dassault Super Etendard being ordered instead.

When Britain and France decided to contribute personnel and material to Operation Desert Storm in 1991, it was inevitable that the Jaguar, which was capable of rapid deployment with minimal support and which could function in relatively primitive conditions, should be included in the Coalition Forces' Order of Battle. In the event, the aircraft performed extremely well.

Specification: SEPECAT Jaguar GR.Mk.1A	
Type:	single-seat tactical support and strike aircraft
Powerplant:	Two 3313kg (7305lb) thrust Rolls-Royce/Turboméca Adour Mk 102 turbofans
Performance:	maximum speed 1593km/h (990mph) at 11,000m (36,090ft); service ceiling 14,000m (46,000ft); combat radius 575km (357 miles) lo-lo-lo
Weights:	empty 7000kg (15,435lb); maximum take-off 15,500kg (34,177lb)
Dimensions:	wing span 8.69m (28ft 6in); length 16.83m (55ft 2in); height 4.89m (16ft 0in); wing area 24.0m² (258.34 sq ft)
Armament:	two 30mm (1.18in) DEFA cannon; five external hardpoints with provision for 4536kg (10,000lb) of underwing stores; two overwing-mounted AIM-9L Sidewinders for self-defence

Dassault/Dornier Alpha Jet

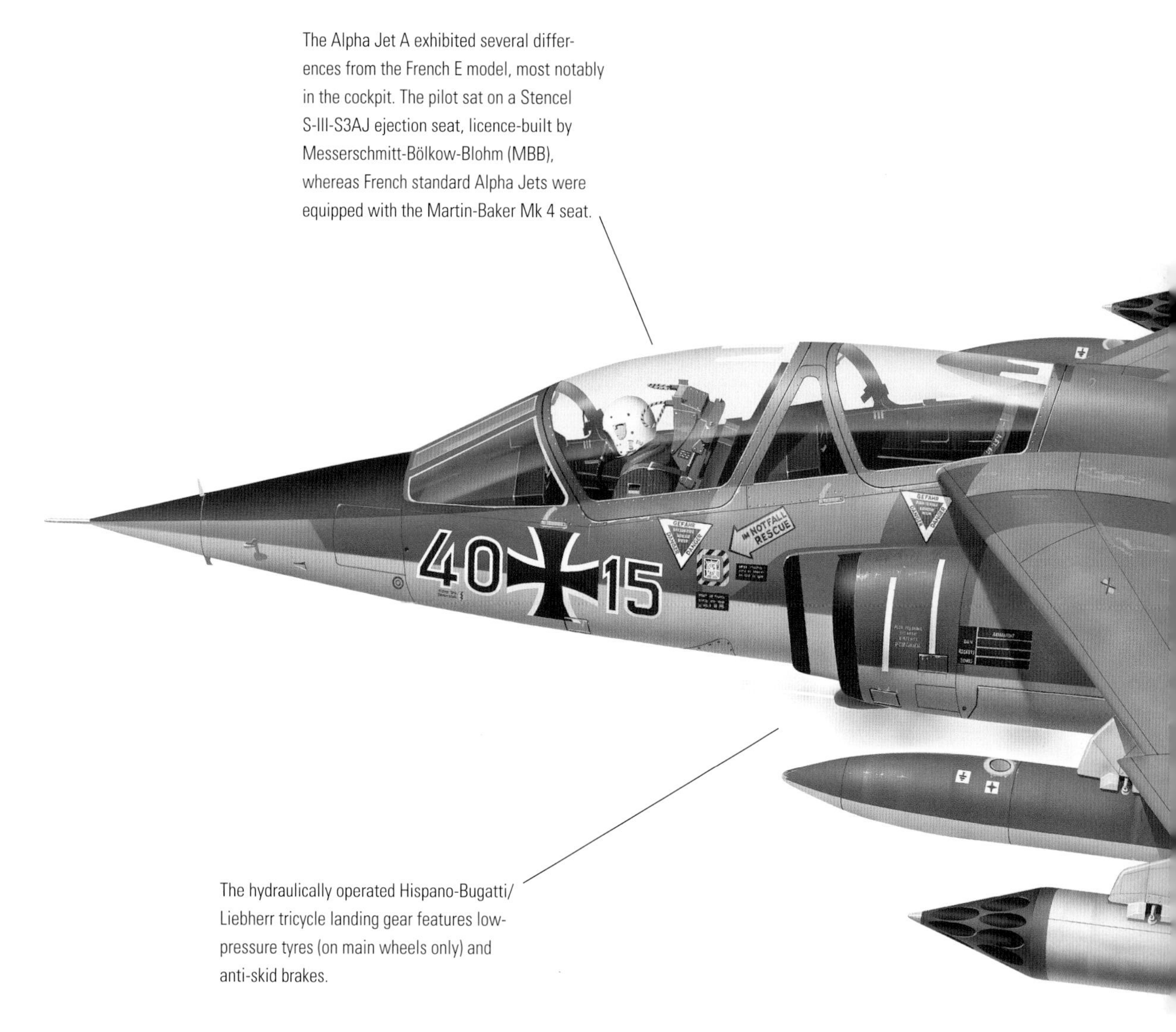

The Alpha Jet A exhibited several differences from the French E model, most notably in the cockpit. The pilot sat on a Stencel S-III-S3AJ ejection seat, licence-built by Messerschmitt-Bölkow-Blohm (MBB), whereas French standard Alpha Jets were equipped with the Martin-Baker Mk 4 seat.

The hydraulically operated Hispano-Bugatti/Liebherr tricycle landing gear features low-pressure tyres (on main wheels only) and anti-skid brakes.

The eagle head badge on the fin of this Alpha Jet is that of JBG 41. This unit was formed as JBG 35 at Husum in 1959, operating Republic F-84F Thunderstreaks. It reformed as JBG 41 in 1961 to operate the Fiat G.91R and was subsequently renamed Leichtenkampgeschwader (light combat wing) 41, reverting to a JBG designation when it received Alpha Jets.

The SNECMA/Turboméca Larzac 04-C6, two of which are installed in the Alpha Jet in nacelles on the fuselage sides, is a turbofan of 1.13 bypass ratio with a two-stage fan, four-stage HP compressor, single-stage HP turbine (having cooled blades) and single-stage LP turbine.

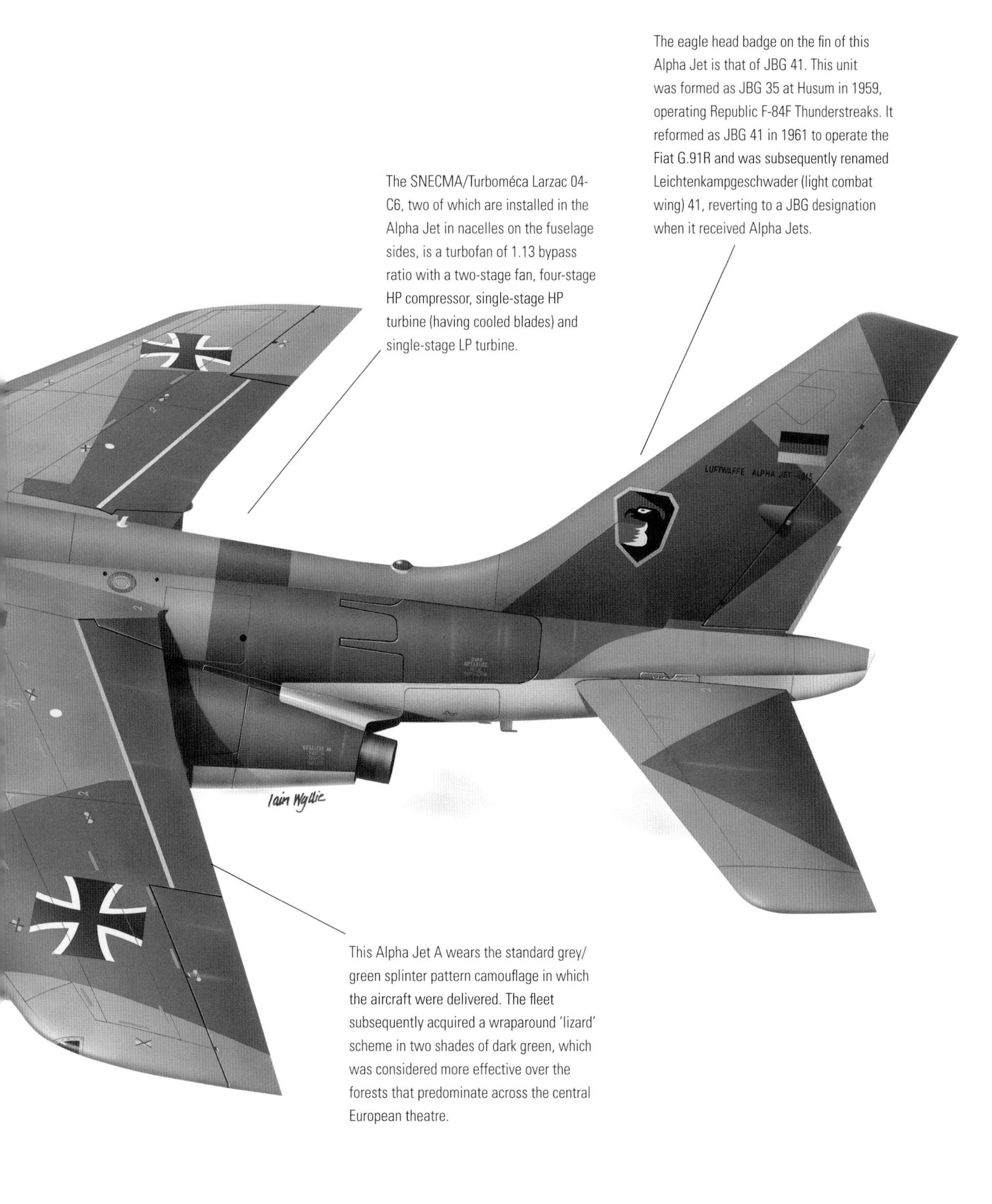

This Alpha Jet A wears the standard grey/green splinter pattern camouflage in which the aircraft were delivered. The fleet subsequently acquired a wraparound 'lizard' scheme in two shades of dark green, which was considered more effective over the forests that predominate across the central European theatre.

In 1968 France and Germany, both of whom had projects for an advanced jet trainer under study, decided to pool their resources and expertise. The result was the excellent Alpha Jet, which, like its rival the British Aerospace Hawk, offered near-fighter performance and attack capability at a significant fraction of the cost of a fully specified combat aircraft. Early in the joint development programme the Germans decided that they had no requirement for a training version of the Alpha Jet, but a need was identified for a light attack aircraft to replace the Fiat G.91R fleet. In February 1972 two prototypes each were ordered by France and Germany, and the type (the French-built version) flew for the first time on 26 October 1973.

Production began some time later, the first trials aircraft being delivered to the French Air Force late in 1977. Re-equipment of French training units started in May 1979, the Alpha Jets replacing the Lockheed-Canadair T-33 and, at a later date, the Mystère IVA single-seat weapons training aircraft. The first German-built Alpha Jet flew for the first time on 12 April 1978. As designated, the two principal versions were Alpha Jet A (A for Appui, or support) light attack aircraft, and the Alpha Jet E (E for Ecole, or School) basic and advanced trainer.

The E model, which also has a light attack capability, was produced for the French Air Force and a number of foreign customers, while deliveries of the A model to the Federal German Luftwaffe began in 1979, 175 being delivered between then and 1983. Most of the Luftwaffe aircraft were assigned a combat role, serving with Jagdbombergeschwader (JBG) 41 at Husum, JBG 43 at Oldenburg and JBG 49 at Fürstenfeldbruck. The latter wing had a dual role, with one Staffel (1/JBG 49) assigned to front-line duties and a second (2/JBG 49) with a conversion training task. This unit also helped to train forward air controllers and provided fast jet experience for prospective Tornado navigators. During the early 1980s the latter task swamped the unit, with the result that many Alpha Jet pilots undertook their conversion with front-line units. A further 18 Alpha Jets were deployed to the Luftwaffe armament practice camp at Beja, Portugal, from where the Luftwaffenkommando Beja undertook weapons training over extensive ranges over land and sea. In time of tension these aircraft would have operated from Leipheim as JBG 44, flown by instructors. By 1993, with tensions in Europe dramatically reduced and a wholesale force reduction in progress, the Alpha Jets were withdrawn from the combat role, and the last trainers were retired in 1997.

Specification: Dassault/Dornier Alpha Jet A	
Type:	single-seat tactical strike aircraft
Powerplant:	two 1350kg (2976lb) Turboméca Larzac 04 turbofans
Performance:	maximum speed 927km/h (576mph); service ceiling 14,000m (45,930ft); low-level mission range 583km (362 miles)
Weights:	empty 3345kg (7375lb); maximum take-off 8000kg (17,640lb)
Dimensions:	wing span 9.11m (29ft 10in); length 13.23m (43ft 5in); height 4.19m (13ft 9in); wing area 17.5m² (188.37 sq ft)
Armament:	one 27mm (1.06in) IWKA-Mauser cannon; five hardpoints with provision for up to 2500kg (5512lb) of stores

Below: **This Alpha Jet E displays the original training colours worn by the aircraft on its entry into service with the Belgian Air Force. Aircraft were later repainted in a two-tone grey camouflage as they returned from overhaul, the orange training bands being retained.**

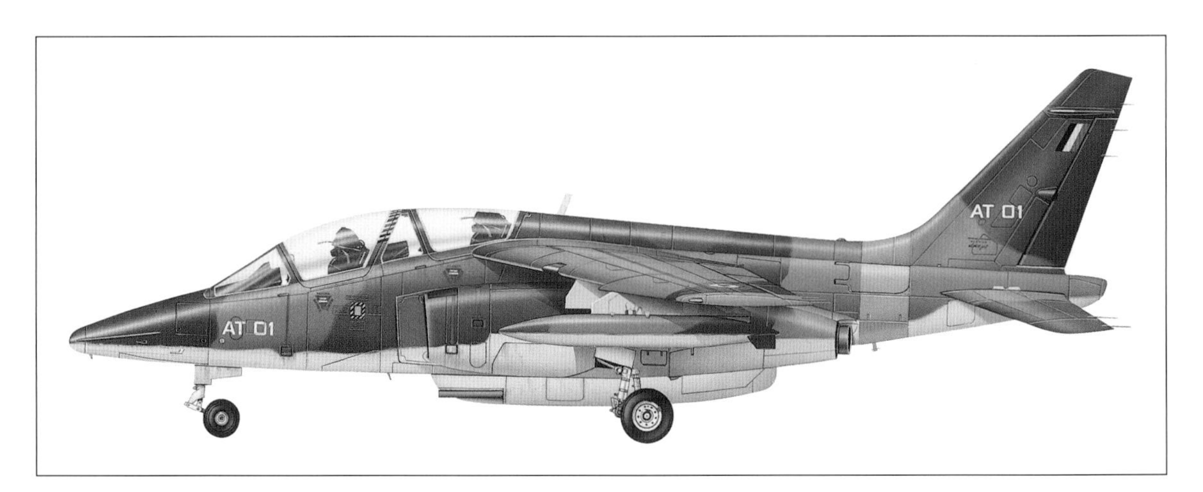

Above: **As this photogaph shows, the Alpha Jet has the ability to operate from stretches of motorway in the event of a war emergency. This technique was pioneered by the world's first jet fighter, the Messerschmitt Me 262, during the closing months of World War II.**

The French Air Force took delivery of 176 aircraft between 1978 and 1985, and export customers have included Belgium, Cameroon, Côte d'Ivoire, Egypt, Morocco, Portugal, Qatar, Thailand and Togo. Some ex-Luftwaffe examples are also used by the UK test agency, DERA. Dassault marketed advanced Alpha Jets under various names, such as the Alpha Jet NGEA (Nouvelle Generation École et Attaque) or Alpha Jet 2. Incorporating 04-C20 engines, a nose-mounted laser rangefinder and upgraded avionics and weapons capability, the NGEA found no customers as such, although the MS2s bought by Egypt and Cameroon were similar. A major upgrade planned for the Luftwaffe's aircraft would have seen the addition of Sidewinder launch rails and the integration of various guided weapons, including Maverick and Harm, but this was cancelled.

Panavia Tornado ADV

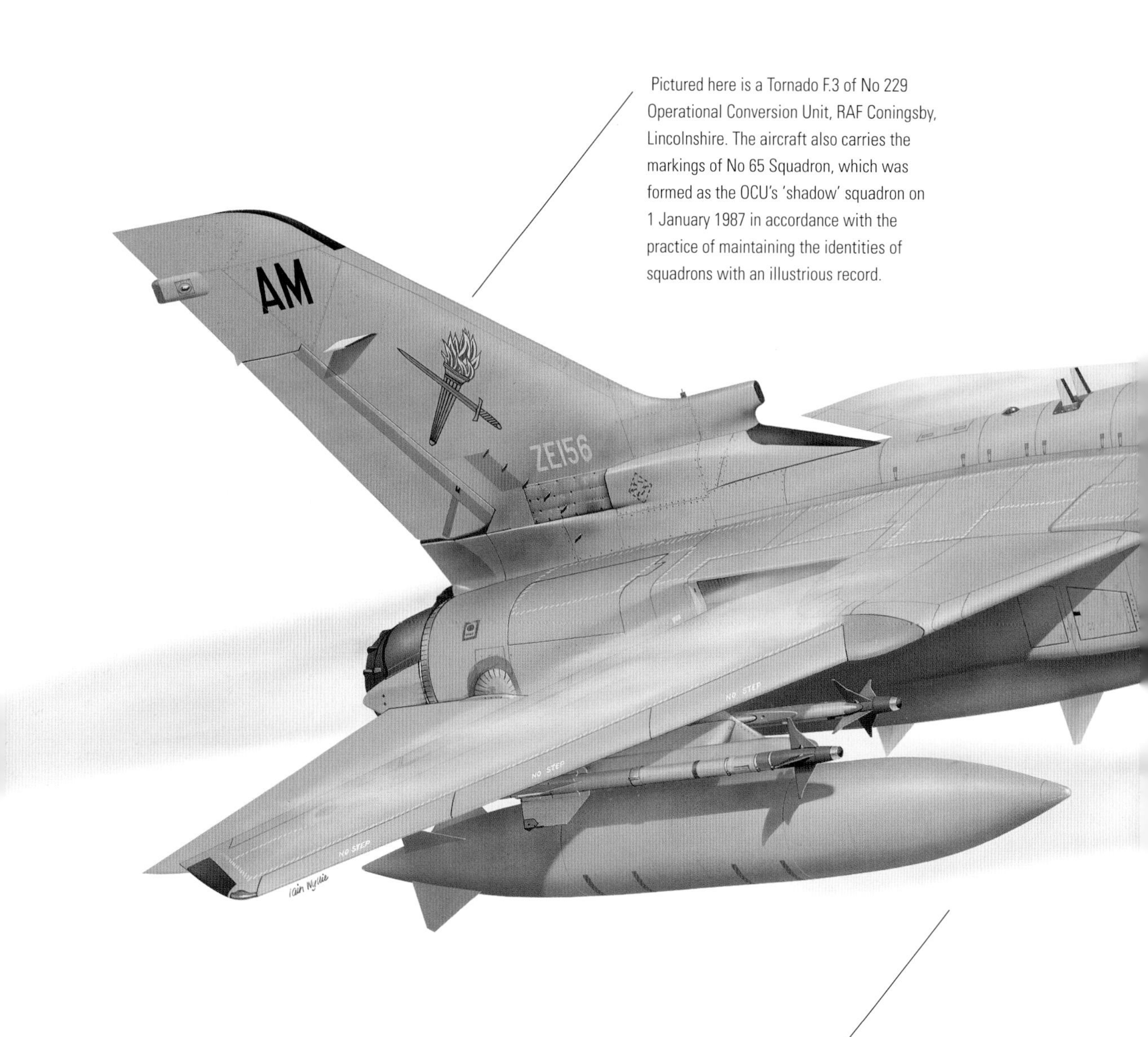

Pictured here is a Tornado F.3 of No 229 Operational Conversion Unit, RAF Coningsby, Lincolnshire. The aircraft also carries the markings of No 65 Squadron, which was formed as the OCU's 'shadow' squadron on 1 January 1987 in accordance with the practice of maintaining the identities of squadrons with an illustrious record.

The overall structural changes involved stretching the fuselage by 136cm (53in); the wing root glove was also given increased sweep, moving the centre of pressure forward to compensate for the resultant change of centre of gravity and to reduce wave drag.

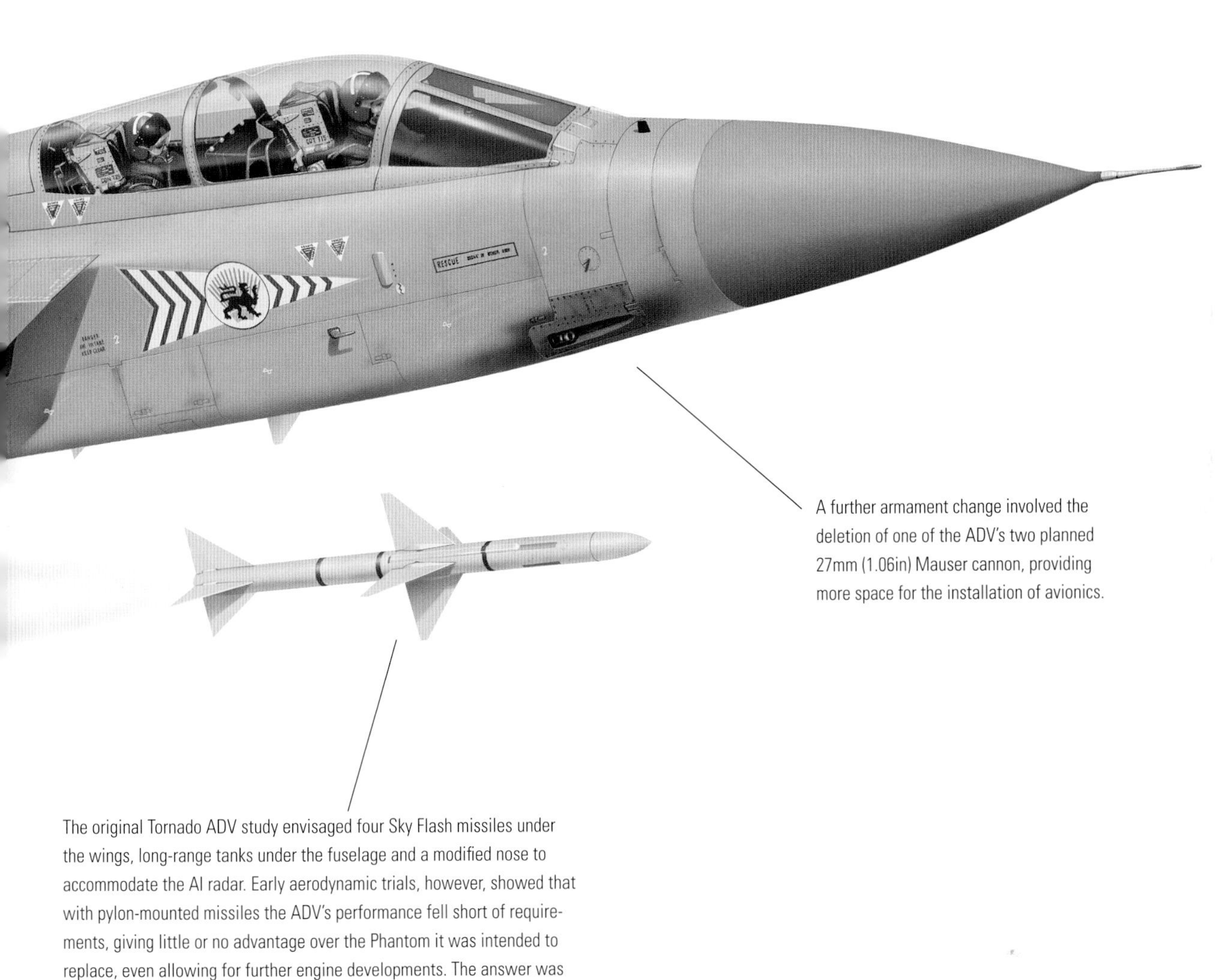

A further armament change involved the deletion of one of the ADV's two planned 27mm (1.06in) Mauser cannon, providing more space for the installation of avionics.

The original Tornado ADV study envisaged four Sky Flash missiles under the wings, long-range tanks under the fuselage and a modified nose to accommodate the AI radar. Early aerodynamic trials, however, showed that with pylon-mounted missiles the ADV's performance fell short of requirements, giving little or no advantage over the Phantom it was intended to replace, even allowing for further engine developments. The answer was to carry the AAMs semi-submerged under the fuselage, providing a low-drag configuration.

In 1971 the UK Ministry of Defence issued Air Staff Target 395, which called for a minimum-change, minimum-cost but effective interceptor to replace the British Aerospace Lightning and the F-4 Phantom in the air defence of the United Kingdom. Primary armament was to be the British Aerospace Dynamics XJ521 Sky Flash medium-range air-to-air missile, and the primary sensor a Marconi Avionics pulse-Doppler radar. The result was the Air Defence Variant (ADV) of the Panavia Tornado interdictor/strike (IDS) aircraft.

The aircraft that eventually emerged was a long-range interceptor, with long on-CAP time, capable of engaging multiple targets in rapid succession, in all weathers and in complex ECM conditions. It was designed to operate with the United Kingdom Air Defence Ground Environment (UKADGE), airborne early warning (AEW) aircraft, tankers and air defence ships, all linked in due course to a secure ECM-resistance data and voice command and control net. The intercept radar selected for the Tornado ADV was the Marconi (later GEC Marconi) Avionics AI24 Foxhunter, development of which began in 1974. The essential requirement was that detection ranges should in no way be limited to target altitude. Look-down capability against low-level targets was the most demanding case, particularly when the interceptor itself was at low altitude. Severe and sophisticated electronic countermeasures also had to be overcome. A great deal of trouble was experienced in the development of this radar, and many modifications had to be made before it was acceptable for service. By the time the first Tornado ADV was ready to fly, late in 1979, the external stores fit had also undergone changes. The four Sky Flash AAMs were now joined by four AIM-9L Sidewinders on underwing stations, and the capacity of each drop-tank increased from 1500 litres to 2250 litres to extend unrefuelled range and time on CAP (combat air patrol).

Three Tornado ADV prototypes were built. All were powered by the Turbo-Union RB.199 Mk103 turbofan, which was also to power the initial production batch of Tornado F.2s for the RAF. These aircraft also featured manually controlled wing sweep, which would be automatic on later production aircraft. The problems with the Foxhunter were still far from resolved when the first Tornado F.2s were delivered to No 229 Operational Conversion Unit (OCU) at RAF Coningsby, Lincolnshire, in November 1984. The first 18 aircraft were all powered by Mk103 engines; aircraft after that had the more powerful Mk104, which combined a 360mm reheat extension with a Lucas Aerospace digital electronic engine control unit (DECU). These later aircraft, designated Tornado F.3 – the definitive production version of the design – also featured the full armament of four Sky Flash and four AIM-9Ls, auto wing sweep, and auto manoeuvre devices with the slats and flaps deploying as a function of angle of attack and wing sweep.

It was not until 1986 that the first modified AI24 Foxhunter radars were delivered for installation in the OCU aircraft, the necessary modifications having cost an additional £250 million. The first squadron, No 29, formed at RAF Coningsby in May 1987 and was declared operational at the end of November. The aircraft eventually armed seven squadrons in addition to No 229 OCU (which became No 56 Reserve Squadron on 1 July 1992).

Below: **The German Navy's Tornado IDS wing, *Marineflieger-geschwader* (MFG) 2, remains an important component of NATO's air defences, and has a formidable anti-shipping capability. This aircraft is armed with Kormoran 2, a digital version of the missile.**

Above: **A Tornado ADV shows off its four Sky Flash air-to-air missiles on their under-fuselage stations. It was originally intended to fit the missiles under the wings, before it was discovered that carrying them semi-recessed under the fuselage was a better option.**

Normal air defence operations with the Tornado F.3 involve what is known as a 'heavy combat fit', which means four Sky Flash, four Sidewinders and no external tanks. CAP fit with the two tanks is reserved specifically for long-range sorties. A good example of what the F.3 can achieve without the long-range tank was given on 10 September 1988, when two aircraft of No 5 Squadron were scrambled from RAF Coningsby to intercept a pair of Tupolev Tu-95 Bear-D maritime radar reconnaissance aircraft over the Norwegian Sea. A VC-10 tanker was scrambled from RAF Leuchars to rendezvous with the Tornadoes, which carried out the intercept successfully.

The Tornado ADV also serves with the air forces of Italy and Saudi Arabia.

Specification: Panavia Tornado ADV

Type:	two-seat all-weather air defence aircraft
Powerplant:	two 7493kg (16,522lb) thrust Turbo-Union RB.199-34R Mk104 turbofans
Performance:	maximum speed 2337km/h (1452mph) above 11,000m (36,090ft); service ceiling 21,335m (70,000ft); intercept radius about 1853km (1000 nautical miles)
Weights:	empty 14,501kg (31,979lb); maximum take-off 27,987kg (61,700lb)
Dimensions:	wing span 13.91m (45ft 8in) spread and 8.60m (28ft 2in) swept; length 18.68m (61ft 3in); height 5.95m (19ft 6in); wing area 26.60m^2 (286.3 sq ft)
Armament:	two 27mm (1.06in) IWKA-Mauser cannon; six external hardpoints with provision for up to 5806kg (12,800lb) of stores, including Sky Flash medium-range AAMs and AIM-9L Sidewinder short-range AAMs, and drop-tanks

McDonnell Douglas/ British Aerospace Harrier II

Unusually for a Western tactical aircraft, the AV-8B(NA) is equipped with chaff/flare launchers on the upper surfaces of the rear fuselage, each side of the ram air inlet.

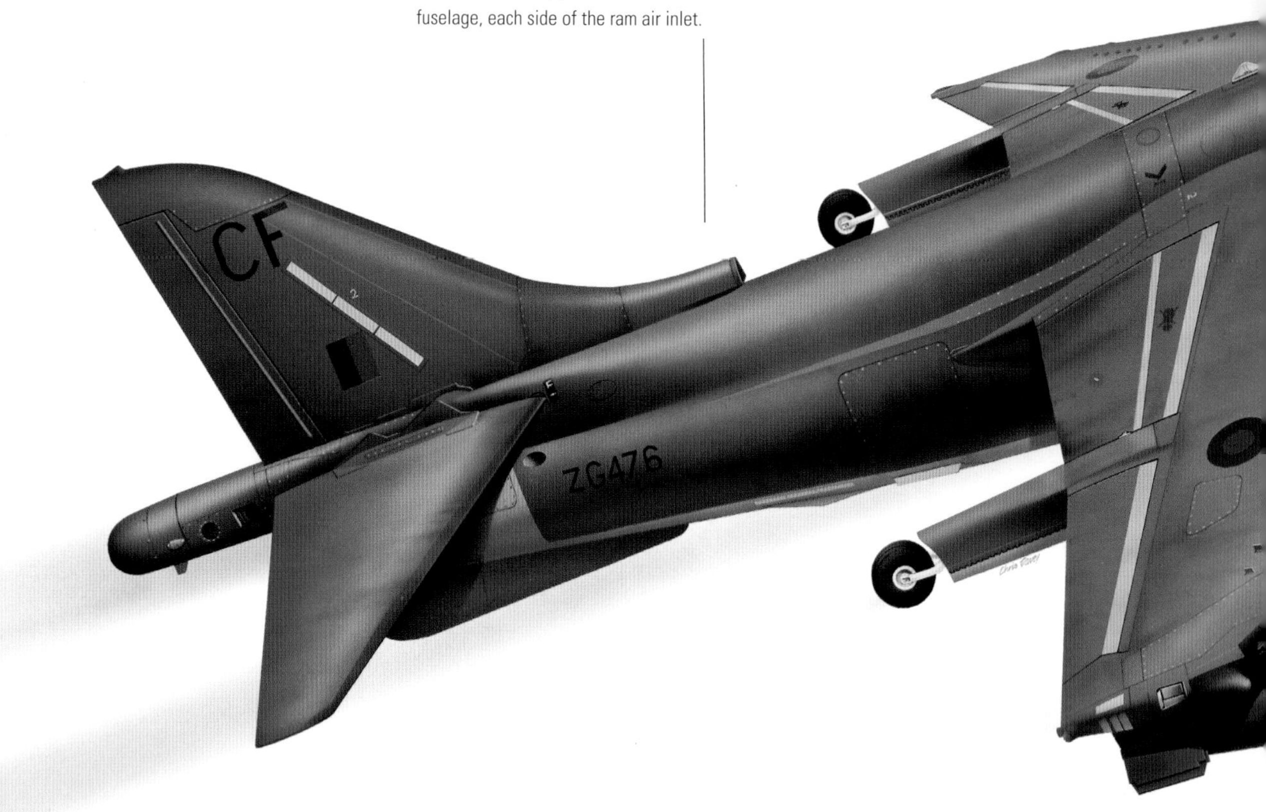

Leading-edge root extensions (LERX) are fitted to enhance the Harrier's air combat agility by improving the turn rate, while longitudinal fences (LIDS, or Lift Improvement Devices) are incorporated beneath the fuselage and on the gun pods to capture ground-reflected jets in vertical take-off and landing, to give a much bigger ground cushion and to reduce hot gas recirculation.

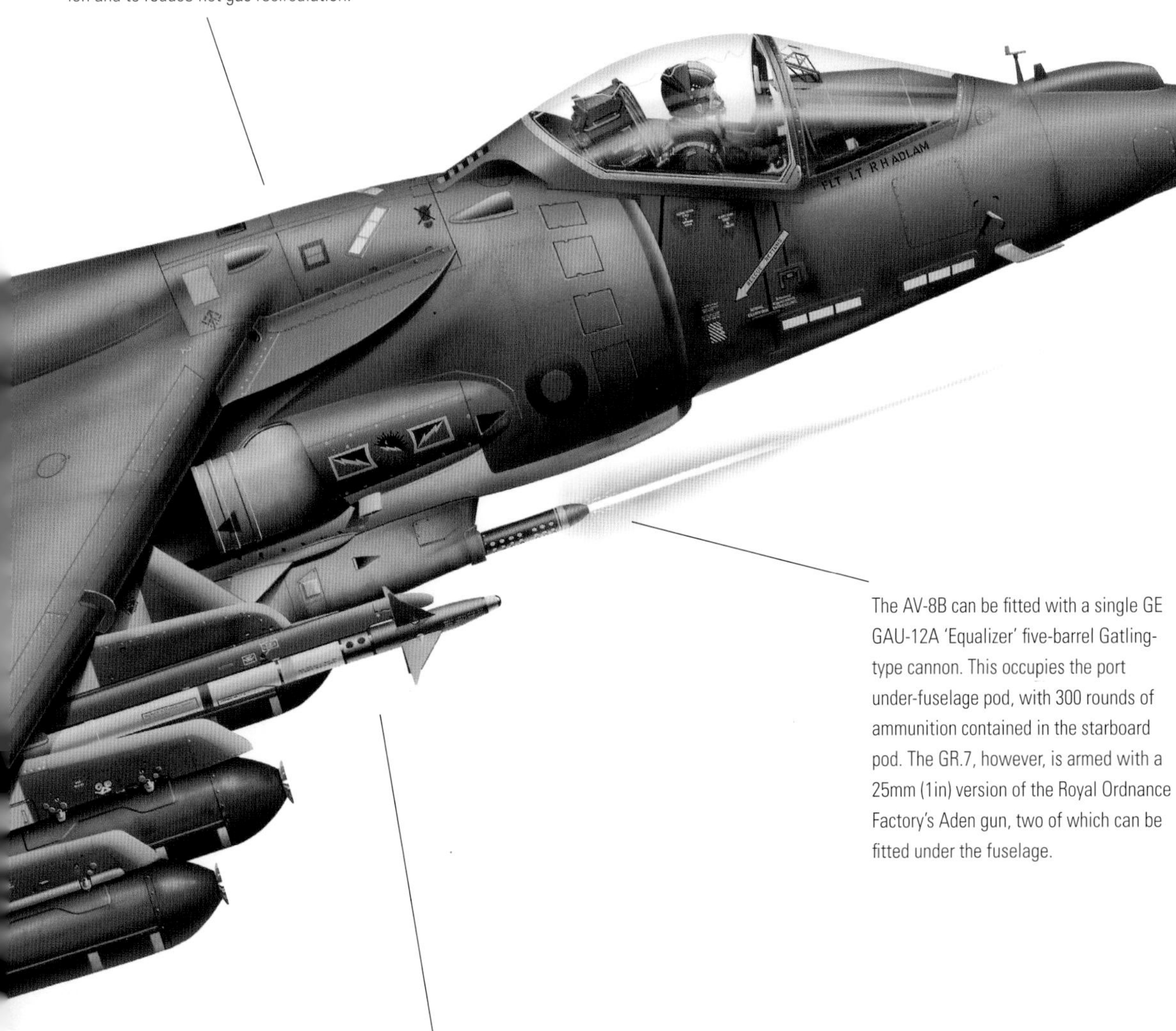

The AV-8B can be fitted with a single GE GAU-12A 'Equalizer' five-barrel Gatling-type cannon. This occupies the port under-fuselage pod, with 300 rounds of ammunition contained in the starboard pod. The GR.7, however, is armed with a 25mm (1in) version of the Royal Ordnance Factory's Aden gun, two of which can be fitted under the fuselage.

The aircraft seen here, ZG476, is a Harrier GR.Mk.7 of No 4 Squadron, one of two Harrier GR.7 units formerly based at Gutersloh, Germany. After a spell at Laarbruch, the squadron returned to the UK (RAF Cottesmore) in 1999.

Above: **The post-Cold War world has seen carrier operations become a vital part of the Harrier force's work. RAF Harriers and Royal Navy Sea Harriers are being combined in a new 'Joint Force' aboard the UK's aircraft carriers.**

A key element in modern offensive battlefield support is the short take-off, vertical landing (STOVL) aircraft, epitomized by the British Aerospace/McDonnell Douglas Harrier. One of the most important and certainly the most revolutionary combat aircraft to emerge during the post-war years, the Harrier V/STOL tactical fighter-bomber began its career as a private venture in 1957 following discussions between Hawker Aircraft Ltd and Bristol Aero-engines Ltd, designers of the BS53 Pegasus turbofan engine. Development of this powerplant, which featured two pairs of connected rotating nozzles, one pair to provide jet lift, was partly financed with American funds, and in 1959–60 the Ministry of Aviation ordered two prototypes and four development aircraft under the designation P.1127. The first prototype made its first tethered hovering flight on 21 October 1960 and began conventional flight trials on 13 March 1961. In 1962 Britain, the United States and West Germany announced a joint order for nine Kestrels, as the aircraft was now known, for evaluation by a tripartite handling squadron at RAF West Raynham in 1965. Six of these aircraft were subsequently shipped to the USA for further trials. In its single-seat close support and tactical reconnaissance version, the aircraft was ordered into production for the RAF as the Harrier GR.Mk.1, the first of an initial order of 77 machines flying on 28 December 1967. On 1 April 1969 the Harrier entered service with the Harrier OCU at RAF Wittering, and the type subsequently equipped No 1 Squadron at Wittering and Nos 3, 4 and 20 Squadrons in Germany.

Left: **Two US Marine Corps AV-8B Harrier IIs over a barren desert landscape. This photograph gives a good impression of the Harrier's high-set cockpit, and the consequent excellent visibility enjoyed by the pilot. Note the wing leading-edge root extensions.**

Although it was the British who were responsible for the early development of this remarkable aircraft, it was the US Marine Corps who identified the need to upgrade its original version, the AV-8A. The Harrier used 1950s technology in airframe design and construction and in systems, and by the 1970s, despite systems updates, this was restricting the further development of the aircraft's potential. In developing the USMC's new Harrier variant the basic design concept was retained, but new technologies and avionics were fully exploited. One of the major improvements was a new wing, with a carbon fibre composite structure, a super-critical aerofoil and a greater area and span. The wing has large slotted flaps linked with nozzle deflection at short take-off unstick to improve control precision and increase lift. Leading-edge root extensions (LERX) are fitted to enhance the aircraft's air combat agility by improving the turn rate, while longitudinal fences (LIDS, or Lift Improvement Devices) are incorporated beneath the fuselage and on the gun pods.

A prototype YAV-8B Harrier II first flew in November 1978, followed by the first development aircraft in November 1981, and production deliveries to the USMC began in 1983.

Right: **An AV-8B Harrier II of Marine Corps Squadron VMA-331 releasing a pair of retarded bombs over a desert practice range. The Harriers unique V/STOL capability means that it can act in support of ground forces within minutes of a call being received.**

The first production AV-8B was handed over to Training Squadron VMAT-203 at Cherry Point, North Carolina, on 16 January 1984, the aircraft making its acceptance check flight four days later. VMAT-203 began training pilots exclusively for the AV-8B in the spring of 1985, and 170 had completed their conversion course by the end of 1986. Operational Harrier pilots were assigned to Marine Air Group (MAG) 32, the first tactical squadron (VMA 331) reaching initial operational capability (IOC) with the first batch of 12 aircraft early in 1985. The squadron's strength had risen to 15 in the autumn of 1986 and had reached the full complement of 20 by March 1987. The second AV-8B tactical squadron, VMA-231, achieved IOC in July 1986 with 15 aircraft, followed by a third squadron, VMA-457, at the end of 1986. The fourth squadron to equip was the first of the West Coast units, VMA-513, which had stood down as the last of the Marine Corps's AV-8A squadron in August 1986.

Delivery of the RAF's equivalent, the Harrier GR5, began in 1987; production GR5s were later converted to GR7 standard. This version, generally similar to the USMC's night-attack AV-8B, has FLIR, a digital moving map display, night vision goggles for the pilot and a modified head-up display. The Spanish Navy also operated the AV-8B, delivered from October 1987. The survivors of an earlier batch of AV-8As were sold to Thailand in 1996.

Specification: McDonnell Douglas/BAe AV-8B Harrier II	
Type:	single-seat V/STOL close support aircraft
Powerplant:	one 10,796kg (23,800lb) thrust Rolls-Royce F402-RR-408 vectored thrust turbofan
Performance:	maximum speed 1065km/h (661mph) at sea level; service ceiling 15,240m (50,000ft); combat radius 277km (172 miles) with 2722kg (6000lb) payload
Weights:	empty 5936kg (13,088lb); maximum take-off 14,060kg (31,000lb)
Dimensions:	wing span 9.25m (30ft 4in); length 14.12m (46ft 4in); height 3.55m (11ft 8in); wing area 21.37m^2 (230 sq ft)
Armament:	one 25mm GAU-12A cannon; six external hardpoints with provision for up to 7711kg (17,000lb) or 3175kg (7000lb) of stores (short and vertical take-off respectively)

WH
VMA-542

Eurofighter Typhoon

The foreplane/delta configuration is intentionally aerodynamically unstable, which provides a high level of agility (particularly at supersonic speeds), low drag and enhanced lift. The pilot controls the aircraft through a computerized digital fly-by-wire system that provides artificial stabilization and gust elevation to give good control characteristics throughout the flight envelope.

The Typhoon is equipped with a CAPTOR (ECR90) multi-mode X-band pulse-Doppler radar, developed by the Euroradar consortium. The multi-mode radar has three processing channels; the third channel is used for jammer classification, interference blanking and sidelobe nulling.

The pilot's control system is a voice-throttle-and-stick system (VTAS). The stick and throttle tops house 24 fingertip controls for sensor and weapon control, defence aids management, and in-flight handling. The direct voice output allows the pilot to carry out mode selection and data entry procedures using voice command.

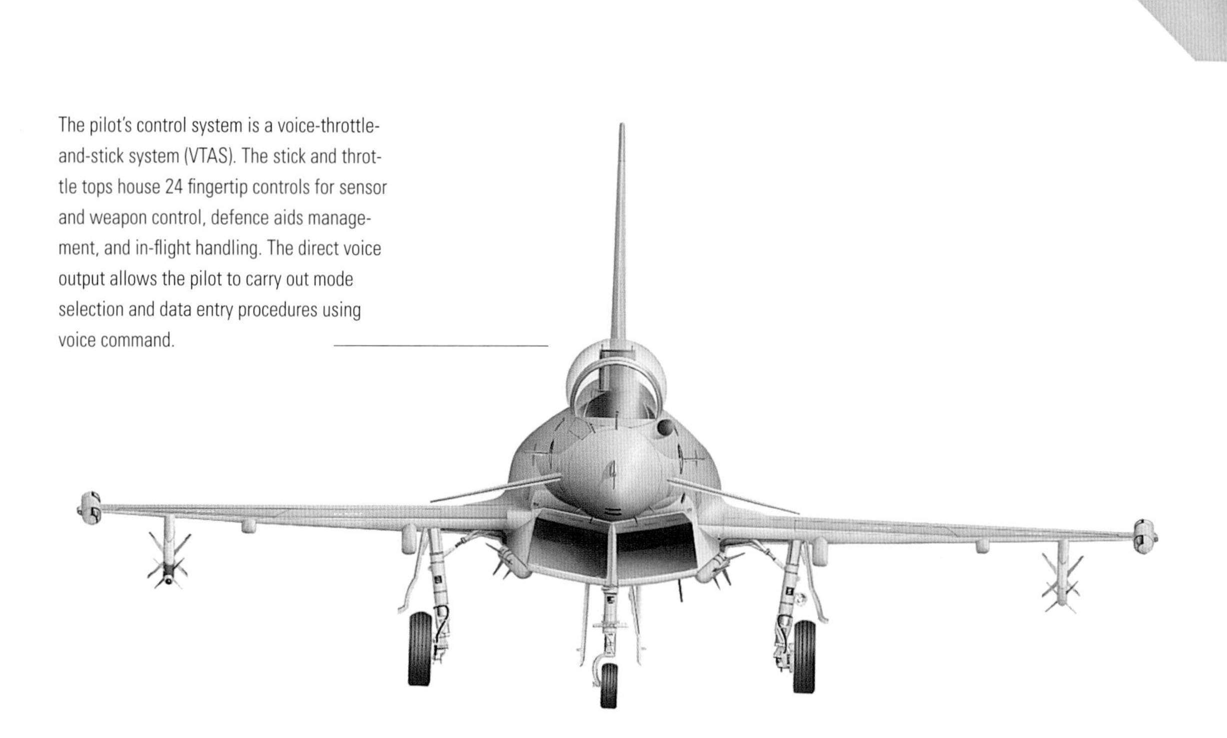

For air-to-air combat, the standard weapon configuration is four beyond-visual-range air-to-air missiles (BVRAAM) on semi-recessed fuselage stations, and two ASRAAM short-range AAMs on the outer pylons. A mix of up to ten medium- and short-range missiles can be carried.

Above: **During 2000, ZH558 (DA.2), the first British-built Eurofighter prototype, was painted black to make it more conspicuous during envelope expansion and handling trials. The aircraft is seen here over Blackpool, not far from its base at Warton, Lancashire.**

In October 1981 the Royal Air Force Operational Requirements Branch began planning for a next-generation fighter to replace the F-4 Phantom in the air defence role, and the Jaguar in the offensive support role. The need crystallized in Air Staff Requirement (Air) 414, which specified a short-range highly agile air defence/offensive support aircraft. The European Fighter Aircraft programme was the project that met this requirement. An outline staff target for a common European fighter aircraft was issued in December 1983 by the air chiefs of staff of France, Germany, Italy, Spain and the UK; the initial feasibility study was completed in July 1984 but France withdrew from the project a year later. A definitive European Staff Requirement (Development), giving operational requirements in greater detail, was issued in September 1987, and the main engine and weapon system development contracts were signed in November 1988.

To prove the necessary technology for EFA, a contract was awarded in May 1983 to British Aerospace for the development of an agile demonstrator aircraft – not a prototype – under the heading Experimental Aircraft Programme, or EAP. The cost was to be shared between the partner companies of the EFA consortium and the UK

Ministry of Defence (MoD). The EAP demonstrator flew for the first time on 8 August 1986, only three years after the programme was conceived. The task of Eurofighter, as EFA ultimately became known, is to fight effectively throughout the combat spectrum, from engagements beyond visual range down to close-in combat. The technologies that enable it to do this are so advanced, and in some cases so unique, that the role of the EAP aircraft was vital to the Eurofighter project as a whole.

The end of the Cold War led, in 1992, to a reappraisal of the whole programme, with Germany in particular demanding substantial cost reductions. Several low-cost configurations were examined, but only two turned out to be cheaper than the original EFA, and both were inferior to the MiG-29 and Su-27. Finally, in December 1992, the project was re-launched as Eurofighter 2000, the planned in-service entry having now been delayed by three years.

Eurofighter is designed to be a 'pilot's aeroplane', with emphasis on the best possible all-round visibility and comfort during high-g manoeuvres. One major asset is the pilot's head-mounted sight, avoiding the need to pull tight turns to achieve missile lock-on and consequently reducing the risk of g-induced loss of consciousness (G-loc). Pilots also have the advantage of a new, fast-reacting g-suit. These innovations mean that there is no need to rake Eurofighter's ejection seat at more than the conventional 18-degree angle, which is good for visibility. It also means that a centrally positioned control column can be retained. Aircraft with highly raked seats, like the F-16, need a sidestick. The cockpit features colour head-down multifunction

Below: **This view clearly shows the angle of the Eurofighter's canard foreplanes, which enhance controllability during high-g manoeuvres and also improve short field performance. The pilot's head-mounted sight avoids the need to pull tight turns to achieve missile lock-on.**

displays and a wide-angle holographic HUD. Direct-voice input (DVI) controls such items as radio channel changes and map displays, but not safety-critical systems such as undercarriage operation or weapon firing. The cockpit area is relatively lightly armoured, providing protection against light-to-medium calibre AAA; the heavy armour is reserved for the critical systems, the thinking being that it is more important to provide the pilot with additional defensive electronics than extra armour plate.

One of the most advanced and ambitious Eurofighter systems is the Defensive Avionics Sub-System (DASS), which was designed to cope with the multiple and mass threats that would have been a major feature of a war on NATO's central front. The system combines and correlates outputs from Eurofighter's radar warning receiver, laser detectors and other sensors and then automatically triggers the best combination of active and passive defences while warning the pilot of the threat priority.

To engage targets, particularly in the vital beyond-visual-range battle, the aircraft is equipped with the Euroradar ECR90 multi-mode pulse-Doppler radar. This is a development of GEC Ferranti's Blue Vixen radar, which is fitted in the British Aerospace Sea Harrier FRS.2. The ECR90 is designed to minimize pilot workload; radar tracks are presented constantly, analysed, allocated priority or deleted by track-management software. A third-generation coherent radar, the ECR90 benefits from a considerable increase in processing power and has all-aspect detection capability in look-up and look-down modes; it also has covert features to reduce the risk of detection by enemy radar warning receivers.

Eurofighter is powered by two EJ200 high performance turbofan engines developing 13,500lb st (60kN) dry and 20,000lb st (90kN) with reheat. The first two Eurofighter prototypes flew in 1994, followed by several more. The original customer requirement was 250 each for the UK and Germany, 165 for Italy and 100 for Spain. The last of these announced a firm requirement for 87 in January 1994, while Germany and Italy revised their respective needs to 180 and 121, the German order to include at least forty examples of the fighter-bomber version. The UK's order was 232, with options on a further 65. Deliveries to the air forces of all four countries were scheduled to begin in 2001 but, not for the first time, the schedule slipped, this time to 2003. However, the type has broken into the export market with an Austrian order for 35 aircraft.

Although Eurofighter is optimized for the air superiority role, a comprehensive air-to-surface attack capability is incorporated in the basic design. Eurofighter is able to carry out close air support, counter-air, interdiction and anti-ship operations; it will also have a reconnaissance capability. Typically, the aircraft has a lo-lo combat radius of 648km (350nm) and a hi-lo-hi combat radius of 1390km (750nm). Eurofighter has a maximum speed of Mach 2.0.

Left: **Another view of Eurofighter Typhoon DA.1 taxiing out to Manchings runway with test pilot Keith Hartley at the controls. The decision of Austria to order 35 aircraft, taken at Farnborough in 2002, was a major breakthrough into the export market.**

Specification: Eurofighter Typhoon	
Type:	single-seat multi-role combat aircraft
Powerplant:	two 9185kg (20,250lb) Eurojet EJ.200 turbofans
Performance:	maximum speed 2125km/h (1320mph) at 11,000m (36,090ft); service ceiling classified; range classified
Weights:	empty 9750kg (21,500lb); maximum take-off 21,000kg (46,300lb)
Dimensions:	wing span 10.50m (34ft 5in); length 14.50m (47ft 7in); height 4.00m (13ft 1in); wing area 52.4m^2 (564.05 sq ft)
Armament:	one 23mm (0.90in) Mauser cannon; 13 hardpoints for a wide variety of ordnance including AMRAAM, ASRAAM, ASMs, anti-radar missiles, guided and unguided bombs

Sweden

As the Cold War developed in the late 1940s, the proximity of the powerful Soviet Union became a matter of growing concern to Sweden, which in World War II had maintained a shaky neutrality not because of powerful defensive forces, but by maintaining a diplomatic foot in both armed camps.

This concern led to the development of some of the world's most potent and advanced combat aircraft. Sweden, with a limited defence budget, mastered the importance of 'getting it right' without wasteful expenditure. It started with a very basic jet fighter design, the Saab J-21R, and rapidly progressed to a more modern design, the J-29, which had the distinction of being the first swept-wing jet fighter of European design to enter service after World War II. Produced in five different versions, it remained in service until well into the 1960s and was also exported to Austria. Its replacement, the J-35 Draken, was designed from the outset to intercept supersonic bombers at all altitudes and in all weathers, and at the time of its service debut in 1960 formed part of the first fully integrated air defence system in western Europe. Capable of Mach 2, it was in service before Britain's one and only indigenous supersonic interceptor, the English Electric Lightning. In the multi-role field, the A-32 Lansen and the AJ-37 Viggen gave the Royal Swedish Air Force an excellent attack capability, a trend that continues today with the JAS-39 Gripen.

Left: **The SAAB JAS 39 Gripen is an excellent design, with sound aerodynamic qualities. Problems in its early development phase were caused mainly by failures in the fly-by-wire control system, which led to the loss of the prototype in an accident.**

Saab J-35 Draken

The nose-mounted radar of air defence versions of the Draken was replaced in the RF-35 by five OMERA cameras, housed in a glass and metal cone, which slid forward on rails to provide access. An additional forward oblique camera was housed immediately behind the glazed panel at the front of the deep undernose section.

This RF-35 Draken is seen in the markings of the Danish Air Force's Eskadrille 729, based at Karup air base in the late 1980s. Denmark placed its original order for Drakens in the form of 20 single-seat F-35s and three TF-35 trainers; it later ordered 20 RF-35 single-seat tactical reconnaissance aircraft and three RF-35 two-seat reconnaissance trainers, of which this is one example.

Situated on the top of the fuselage behind the cockpit, a prominent blade antenna enabled the crew to communicate externally by means of Very High Frequency (VHF) radio. Another blade antenna was located on the under-fuselage, immediately aft of the nosewheel bay. This allowed Ultra-High Frequency (UHF) transmissions and Tactical Aid to Navigation (TACAN) navigation.

The Saab J-35's distinctive 'double delta' wing planform was first tested on a small research aircraft, the Saab 210, in the early 1950s. It still appears futuristic today, and its aerodynamic advantage was that it combined the low drag of the delta wing with astonishing low-speed manoeuvrability.

Above: **A trio of SAAB J-35 Drakens in flight. When it entered service in 1960, the Draken was a component of the most advanced air defence system in western Europe, being integrated with the STRIL 60 air defence environment and having a collision-course fire control system.**

The road that led to the development of the Saab J-35 Draken began with the J-21R, a jet-powered version of the piston-engined, twin-boom J-21A. The J-21R flew on 10 March 1947 but, owing to many modifications that had to be made to the airframe, production deliveries did not take place until 1949, and an order for 120 aircraft was cut back to 60. After a short career as a fighter the J-21 was converted to the attack role as the A-21R. The type was the only aircraft ever to see first-line service with both piston and jet power. The J-21R was followed into service by the swept-wing Saab J-29. The first of three prototypes flew on 1 September 1948 and the first production model, the J-29A, entered service in 1951. Other variants of the basic design were the J-29B, with increased fuel tankage; the A-29 ground attack version, identical to the J-29 except for underwing ordnance racks; and the S-29C reconnaissance version. In the autumn of 1946, Saab had also begun design studies of a new turbojet-powered attack aircraft for the Swedish Air Force, and two years later the Swedish Air Board authorized the construction of a prototype under the designation P1150. This aircraft, now known as the A-32 Lansen (Lance), flew for the first time on 3 November 1952, powered by a Rolls-Royce Avon RA7R turbojet. Three more prototypes were built, and one of these exceeded Mach 1 in a shallow dive on 25 October 1953. The A-32A attack variant was followed by the J-32B all-weather fighter, which first flew in January 1957.

Meanwhile, design work had been proceeding on the next-generation Swedish combat aircraft, the Saab J-35 Draken, which represented a great step forward from anything that had gone before. Designed from the outset to intercept transonic bombers at all altitudes and in all weathers, the Draken was, at the time of its service debut, a component of the finest fully integrated air defence system in western Europe. The first of three prototypes of this unique 'double delta' fighter flew for the first time on 25 October 1955, and the initial production version, the J-35A, entered service early in 1960. The J-35C was a two-seat operational trainer. The J-35D was the first service model to have the more powerful RM6C engine. The major production version of the Draken was the J-35F, which was effectively designed around the Hughes HM-55 Falcon radar-guided air-to-air missile and was fitted with an improved S7B collision-course fire control system, a high capacity datalink system integrating the aircraft with the STRIL 60 air defence environment, an infrared sensor under the nose and PS-01A search and ranging radar. The Saab RF-35 was a reconnaissance version, which continued to serve with Eskadrille 729 of the Royal Danish Air Force until the unit was disbanded as an economy measure and its task was taken over by the F-16s of Esk 726 in January 1994. Total production of the Draken was around 600 aircraft, equipping 17 RSAF squadrons; the type was also exported to Finland and Austria as well as Denmark, which purchased 51 examples in three different versions. Finland bought 49 aircraft, and also licence-built a further 12 at the Valmet factory, while 24 were purchased by Austria. The latter's Drakens were still in service in 2002 with Nos 1 and 2 Squadrons at Zeltweg and Graz respectively. The Draken was the first fully supersonic aircraft in western Europe to be deployed operationally.

Specification: Saab J-35J Draken

Type:	single-seat all-weather interceptor
Powerplant:	one 7830kg (17,265lb) Svenska Flygmotor RM6C (licence-built Rolls-Royce Avon 300 series) turbojet
Performance:	maximum speed 2125km/h (1320mph) at altitude; service ceiling 20,000m (65,600ft); range 3250km (2020 miles) with maximum fuel
Weights:	empty 7425kg (16,372lb); maximum take-off 16,000kg (35,280lb)
Dimensions:	wing span 9.40m (30ft 10in); length 15.40m (50ft 6in); height 3.90m (12ft 9in); wing area 49.20m² (526.6 sq ft)
Armament:	one 30mm (1.18in) Aden cannon; four AAMs (two radar-homing Rb 27 and two IR-homing Falcon, or four Rb 24 Sidewinder); up to 4082kg (9000lb) of bombs on attack mission

Below: **This J-35J was specially painted to celebrate a squadron anniversary, displaying the 3rd Division's swordfish badge above and below each wing as well as on the tail fin. The machine was then used as the Division's aerobatic display aircraft.**

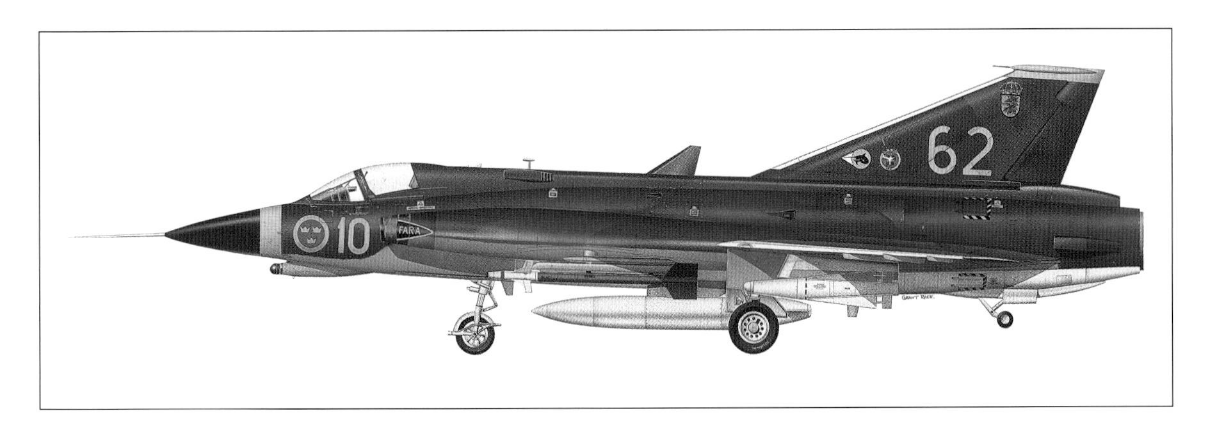

Saab JA-37 Viggen

The Viggen's wraparound single-piece windscreen gives the pilot an excellent view forward, and is strengthened to withstand bird-strikes at high speed.

The oval-shaped lateral air intakes stand well away from the fuselage to avoid the ingestion of sluggish boundary layer air, and are of plain, fixed type, since high supersonic speeds, where a more complex variable intake would be required, are of secondary importance.

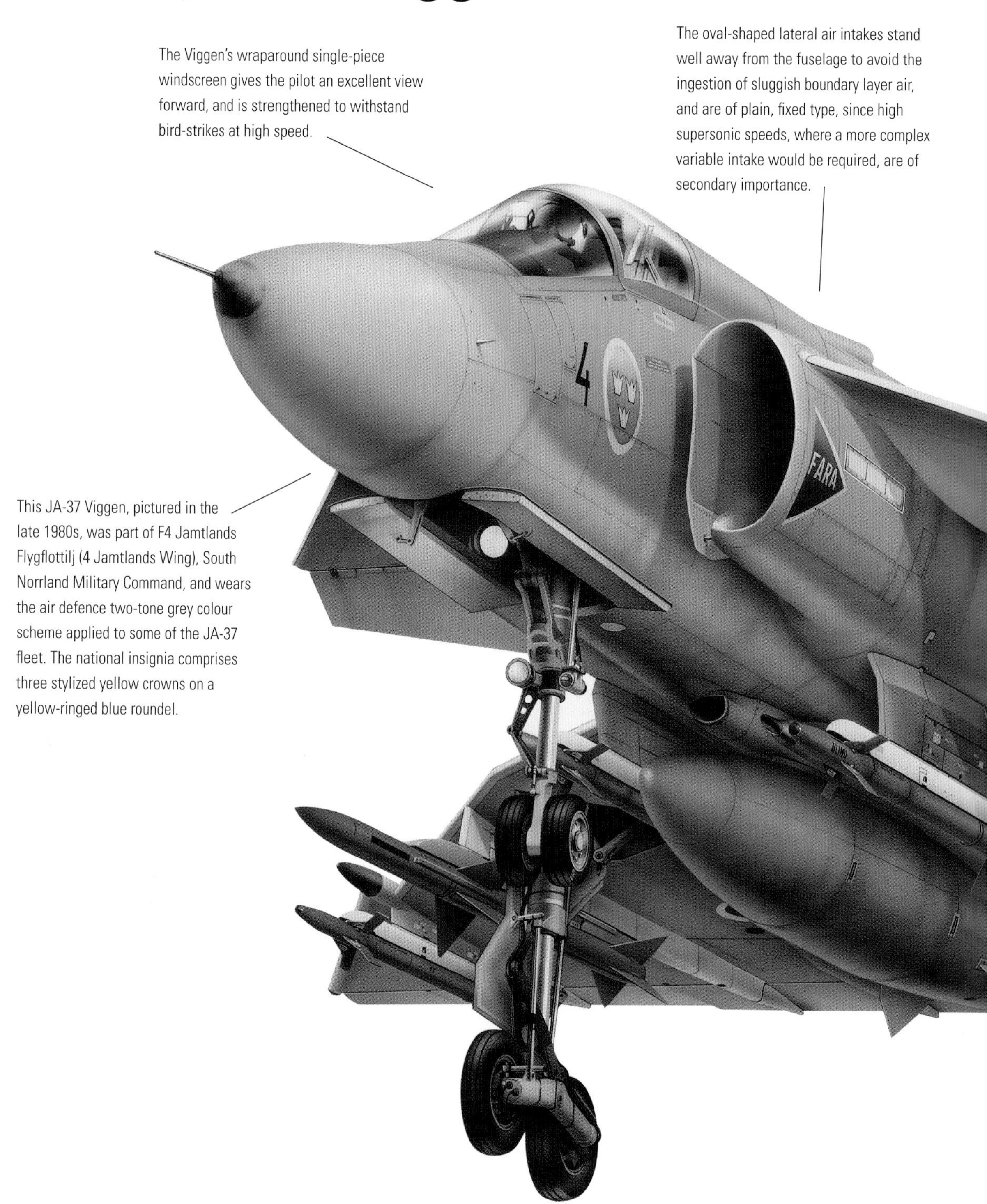

This JA-37 Viggen, pictured in the late 1980s, was part of F4 Jamtlands Flygflottilj (4 Jamtlands Wing), South Norrland Military Command, and wears the air defence two-tone grey colour scheme applied to some of the JA-37 fleet. The national insignia comprises three stylized yellow crowns on a yellow-ringed blue roundel.

The fin can be folded down to port, reducing aircraft height and facilitating storage in Sweden's network of underground hangars.
The wing incorporates hydraulically activated two-section elevons on the trailing edge; the leading edge has compound sweep and is extended forward on the outer sections, outboard of the prominent bullet fairings that accommodate RW antennae.
51
51
Chris Davey

Until the debut of the Panavia Tornado, it may be argued that the Saab Viggen was the most advanced combat aircraft ever produced in Europe, possessing a far more advanced radar, greater speed range and a more comprehensive avionics fit than its contemporaries. Certainly one of the most potent combat aircraft of the 1970s, the Saab 37 Viggen (Thunderbolt) was designed to carry out the four roles of attack, interception, reconnaissance and training. Like the earlier J-35 Draken, it was fully integrated into the STRIL 60 air defence control system. Powered by a Swedish version of the Pratt & Whitney JT8D turbofan engine, with a powerful Swedish-developed afterburner, the aircraft had excellent acceleration and climb performance. Part of the requirement was that it should be capable of operating from sections of Swedish motorways. The first of seven prototypes flew for the first time on 8 February 1967, followed by the first production AJ-37 single-seat all-weather attack variant in February 1971.

Deliveries of the first of 110 AJ-37s to the Royal Swedish Air Force began in June that year. The JA-37 interceptor

Below: **A trio of SAAB Viggens in flight. Although it could hardly be described as aesthetically appealing, the Viggen was a highly effective multi-role combat aircraft that would undoubtedly have given an excellent account of itself if Sweden's neutrality had been breached.**

version of the Viggen, 149 of which were built, replaced the J-35F Draken; the most fundamental difference between this variant and its predecessor lay in the modifications under the skin, rather than in external variations. The JA-37 replaced the PS-37/A with the Ericsson PS-46A medium-PRF multi-mode X-band pulse-Doppler look-down/shoot-down radar, which has four air-to-air modes and a look-down range in excess of 48km (30 miles). The dedicated interceptor JA-37 Viggen was armed with up to six air-to-air missiles for beyond-visual-range combat, the standard weapon being the British Aerospace Dynamics Sky Flash. For short-range work, the Rb 74 (AIM-9L) IR-homing Sidewinder was used. The JA-37 also introduced an integral high-velocity 30mm (1.18in) Oerlikon KCA revolver cannon (offset to port) with 150 rounds mounted in a ventral pack, and aimed by a new radar-based weapons suite. The JA-37 Mod D upgrade added AIM-120 AMRAAM capability to the 'fighter Viggen'.

For ground attack missions, the AJ and AJS Viggen used a variety of air-to-ground missiles, including indigenous weapons like the Rb 04 and Rb 05, and foreign-designed weapons like the AGM-65 Maverick. The main long-range anti-shipping missile in use by the Viggen has been the RB-15F, originally designed as a shipborne weapon, which used the same basic airframe as the Rb 04, albeit with new cruciform tail-fins replacing the old 'wings and endplates'. It is powered by a French Microturbo TRI-60-3 turbofan engine, which gives a range in excess of 90km (55 miles), while the combination of inertial and active radar guidance allows 'fire and forget' operation. Other weapons carried in the air-to-ground role include up to 16 M63 FFV 120kg (265lb) bombs. A favourite weapons option against soft targets was the Bofors M70 rocket pod, containing six

Above: **A SAAB Viggen of F13, which flew a mixture of SF 37 and SH 37 reconnaissance variants as well as JA 37 fighters. F13 disbanded at a later date and its aircraft were allocated to 1 Attack/Spaningsdivision, which disbanded in turn in April 2000.**

135mm (5.3in) rocket projectiles. Up to four of these could be carried, or alternatively the Viggen could carry a pair of 30mm (1.18in) cannon pods.

The SF-37 (26 delivered) was a single-seat armed photo reconnaissance variant; and the SH-37 (26 delivered) was an all-weather maritime reconnaissance version, replacing the S-32C Lansen. The SK-37 (18 delivered) was a tandem two-seat trainer, retaining a secondary attack role. Some Viggens were expected to remain in service until 2010.

Specification: Saab SF-37 Viggen

Type:	single-seat all-weather photo-reconnaissance aircraft
Powerplant:	one 11,797kg (26,015lb) thrust Volvo Flygmotor RM8 turbofan
Performance:	maximum speed 2124km/h (1320mph) at altitude; service ceiling 18,290m (60,000ft); combat radius 1000km (621 miles) hi-lo-hi, with external stores
Weights:	empty 11,800kg (26,015lb); maximum take-off 20,500kg (45,202lb)
Dimensions:	wing span 10.60m (34ft 9in); length 16.30m (53ft 5in); height 5.60m (18ft 4in); wing area 46m² (495.16 sq ft)
Armament:	in secondary attack role, seven external hardpoints with provision for 6000kg (13,230lb) of stores, including 30mm (1.18in) Aden cannon pods, 135mm (5.3in) rocket pods, AAMs and ASMs

Saab JAS-39A Gripen

Subdued markings, a very low-visibility colour scheme and the Gripen's small size all combine to make it a difficult opponent in close-in dogfighting. Some pilots, however, have noted that the aircraft's holographic head-up display is so large that it can produce distinctive green flashes of sun 'glint' that are sometimes bright enough to betray the JAS-39's position.

This Gripen, which belongs to the Flygvapnet's F7 Wing at Såtenäs in the country's Southern Air Command, is seen in a mixed attack/defence load with two BK90 (DWS 39) glide weapons on the inboard wing pylons and Rb 99 (AIM-120) AMRAAMs on the outboard pylons. AIM-9 Sidewinders are mounted on the wingtip pylons.

In designing the JAS-39, Saab retained the tried and tested aft-mounted delta wing configuration, with swept canard foreplanes, an arrangement that makes for excellent manoeuvrability at all speeds and altitudes.

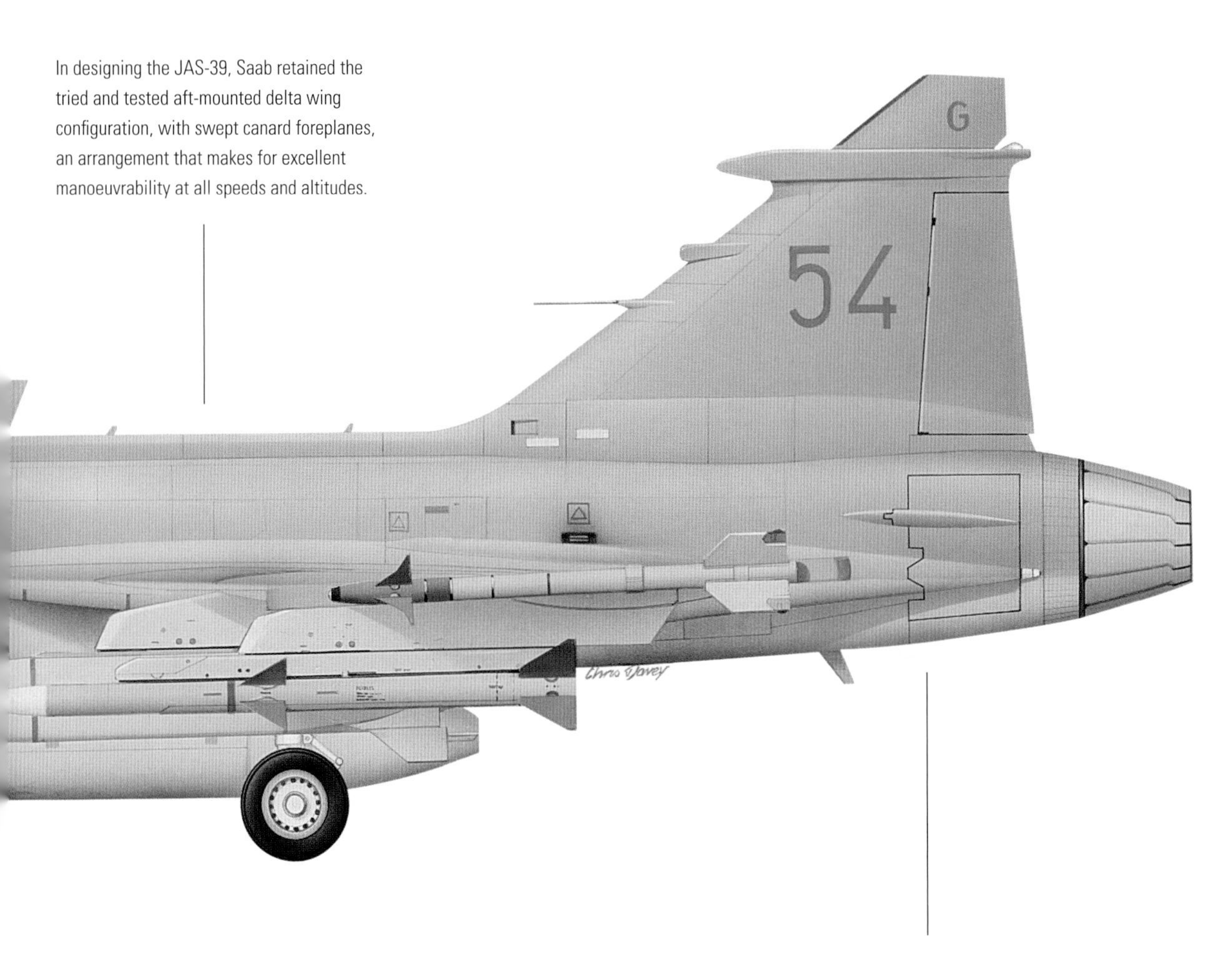

Carrying a full war load, a Gripen can reach 10,000m (33,000ft) in less than two minutes from starting its take-off roll. At low altitudes the aircraft can reach Mach 1.5, although it needs about 30 seconds to accelerate from Mach 0.5 to Mach 1.5. The aircraft has been flown to Mach 2 at altitude.

The Saab JAS-39 Gripen (Griffon) lightweight multi-role fighter was conceived in the 1970s as a replacement for the attack, reconnaissance and interceptor versions of the Viggen. The JAS-39 is a canard delta design with triplex digital fly-by-wire controls, a multi-mode Ericsson pulse-Doppler radar, laser inertial navigation system, wide-angle head-up display and three monochrome head-down displays. The aircraft's Volvo Flygmotor RM12 turbofan (a licence-built General Electric GE F404) is hardened against bird-strike. On 30 June 1982 Saab received a contract for the production of five prototypes and an initial batch of 30 production aircraft, with options for a further 110. The overall go-ahead was confirmed in the second quarter of 1983; test runs of the Volvo Flygmotor RM12 turbofan engine began in January 1985; the selected head-up display (HUD) equipment was first flown in a Viggen test bed in February 1987; and studies of a two-seat JAS-39B version were authorized in July 1989. Meanwhile, the prototype (39-1) had been rolled out on 26 April 1987 and made its first flight on 9 December 1988, but only six sorties were flown in this aircraft before it was lost in a landing accident on 2 February 1989. This led to a revision of the Gripen's advanced fly-by-wire control system, the failure of which had been the cause of the crash. The other Gripen prototypes flew on 4 May 1990 (39-2), 20 December 1990 (39-4), 25 March 1991 (39-3) and 23 October 1991 (39-5). At the end of that year the modified Viggen test bed (37-51) was retired after having made some 250 flights on avionics trials, its task over. A second production batch of 110 aircraft was approved on 3 June 1992; the first production Gripen flew on 10 September 1992 and joined the test programme in place of the first prototype, 39-1. The first production aircraft for the Royal Swedish Air Force (39-102) made its first flight on 4 March 1993 and was handed over to FMV 8 (Försvarets Materielverk 8, a government-controlled test and development organization, rather like Britain's Royal Aircraft Establishment) on 8 June 1993, but this aircraft was lost in a crash on 8 August, resulting in further modifications to the Gripen's flight control software.

The 2000th Gripen sortie was flown on 22 September 1995 by 39-4 and all development work under the original contract had been completed by late 1996, the six development aircraft completing more than 1800 hours' flying in 2300 sorties. The programme included high angle

Right: **Like all Swedish combat aircraft, the Gripen is capable of operating from stretches of motorway and dispersing into forest clearings. It can also be housed in underground hangars carved out of cliffs. The emphasis is on survival for as long as possible.**

20

Left: **The Gripen shows its paces, and its weaponry. The first unit to arm with the type was F7 Wing, and Initial Operational Capability (IOC) was achieved in September 1997. The type has replaced the Viggen in Royal Swedish Air Force Service.**

of attack (AoA) trials and spin trials. By 1996 the Gripen had demonstrated its ability to cruise at Mach 1.08 without reheat. Further trials were conducted with a mock-up flight refuelling probe, the fourth development aircraft carrying out in-flight refuelling link-ups in the course of eight sorties with an RAF VC10 K.Mk.4 between 2 and 17 November 1998.

Orders for the Gripen totalled 140 aircraft, all for the Royal Swedish Air Force. The first unit to arm with the type was F7 Wing at Såtenäs. Maintenance training began in May 1994 at Linköping; conversion training was scheduled to begin in October 1995 but this was postponed until 1996, with the pilot training centre at Såtenäs officially opened in June that year. The Gripen achieved Initial Operational Capability (IOC) in September 1997, following a three-week field exercise by No 2 Squadron of F7, and the last of this unit's JA-37 Viggens was withdrawn in October 1998.

Below: **The Gripen is marketed jointly by SAAB and British Aerospace. The first export customer was South Africa, which in 1999 placed an order for 28 Gripens and 24 BAe Hawk 100s as part of a package, deliveries to begin in 2005 for completion in 2012.**

Specification: Saab JAS-39A Gripen

Type:	single-seat multi-role combat aircraft
Powerplant:	one 8210kg (18,100lb) thrust Volvo Flygmotor RM12 turbofan
Performance:	maximum speed Mach 2 plus; service ceiling classified; range 3250km (2020 miles)
Weights:	empty 6622kg (14,600lb); maximum take-off 12,473kg (27,500lb)
Dimensions:	wing span 8.00m (26ft 3in); length 14.10m (46ft 3in); height 4.70m (15ft 5in)
Armament:	one 27mm (1.06in) Mauser BK27 cannon; six external hardpoints for Sky Flash and Sidewinder AAMs, Maverick ASMs, anti-ship missiles, bombs, cluster bombs, reconnaissance pods, drop-tanks, ECM pods etc

By mid-2000 Swedish Air Force Gripens had flown 16,000 sorties in 12,000 hours, the 100th aircraft being delivered on 12 March 2001.

Despite the fact that Sweden's impact on the military aircraft export market to date has not been spectacular, the Gripen has registered one major success; on 3 December 1999 the South African Air Force announced that Saab and British Aerospace would supply 28 Gripens and 24 Hawk 100s, to be delivered between 2005 and 2012.

USSR/ Russia

In comparison with Western combat types that were making their appearance in the early 1950s, Soviet designs appeared crude, lacking the elegance and refinement that were the hallmarks of a successful aircraft.

But Soviet designers were learning quickly, and before the end of the decade the Soviet Air Force would have at its disposal some of the most effective military aircraft in the world. In 1955, after a slow and uncertain entry into the jet age, the Russian designers were poised to make a great leap forward, and the extent of that leap became apparent to Western observers at the 1955 Tushino Air Display, when the Russians paraded a succession of new types such as the Tupolev Tu-16 and Tu-95 bombers, and the MiG-19, the first Soviet aircraft capable of exceeding the speed of sound in level flight. In the decades that followed the Soviet Union made great strides in developing aircraft to match, or counter, those in the NATO arsenal, including the Sukhoi Su-24 'Fencer', designed to perform roughly the same tasks as the General Dynamics F-111. Despite the continued development and successful export of multi-role aircraft such as the MiG-29 'Fulcrum', the breakup of the Soviet Union after 1990 caused massive disruption to the nation's military aircraft programmes.

Left: **The Su-27M (Su-35) was developed as an advanced multi-role derivative of the Su-27 and was intended for the export market as much as for home service, but in competition with international fighter designs such as the F-16 it has met with no success.**

Ilyushin Il-28 Beagle

The Il-28 was one of the classic examples of Soviet exploitation of British jet engine technology. The Nene engine, probably the ultimate centrifugal-flow turbojet, was a windfall, being copied as the RD-45. The Il-28 was actually powered by an improved version, designated VK-1A (the initials being those of Vladimir Klimov, the engine manufacturer).

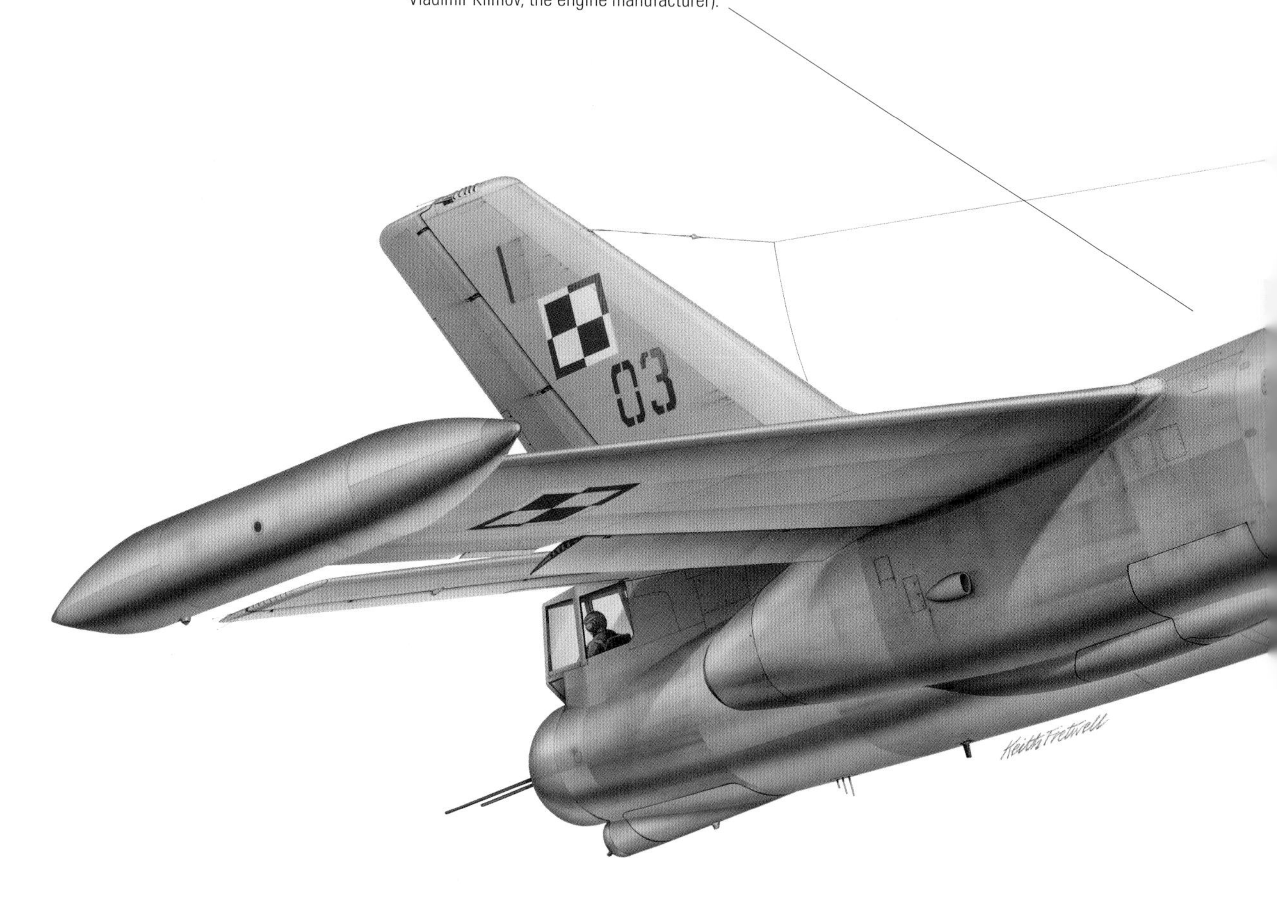

The pilot sat on a KM-1 series ejection seat, under a clear view canopy that hinged to starboard for access. The solid rear fairing incorporated a flush antenna for DF equipment, and an aerial mast for the HF antenna, which ran down from the fin. The use of a fighter-style cockpit allowed the fuselage cross-section to be kept as slim as possible.

When not sitting in his ejection seat just ahead of the pilot's cockpit, the Il-28 navigator could move forward to lie prone in the nose to use the gyro-stabilized OPB-5 optical bomb sight. For this, an optically flat, undistorted panel was set into the underside of the heavily framed nose-cone.

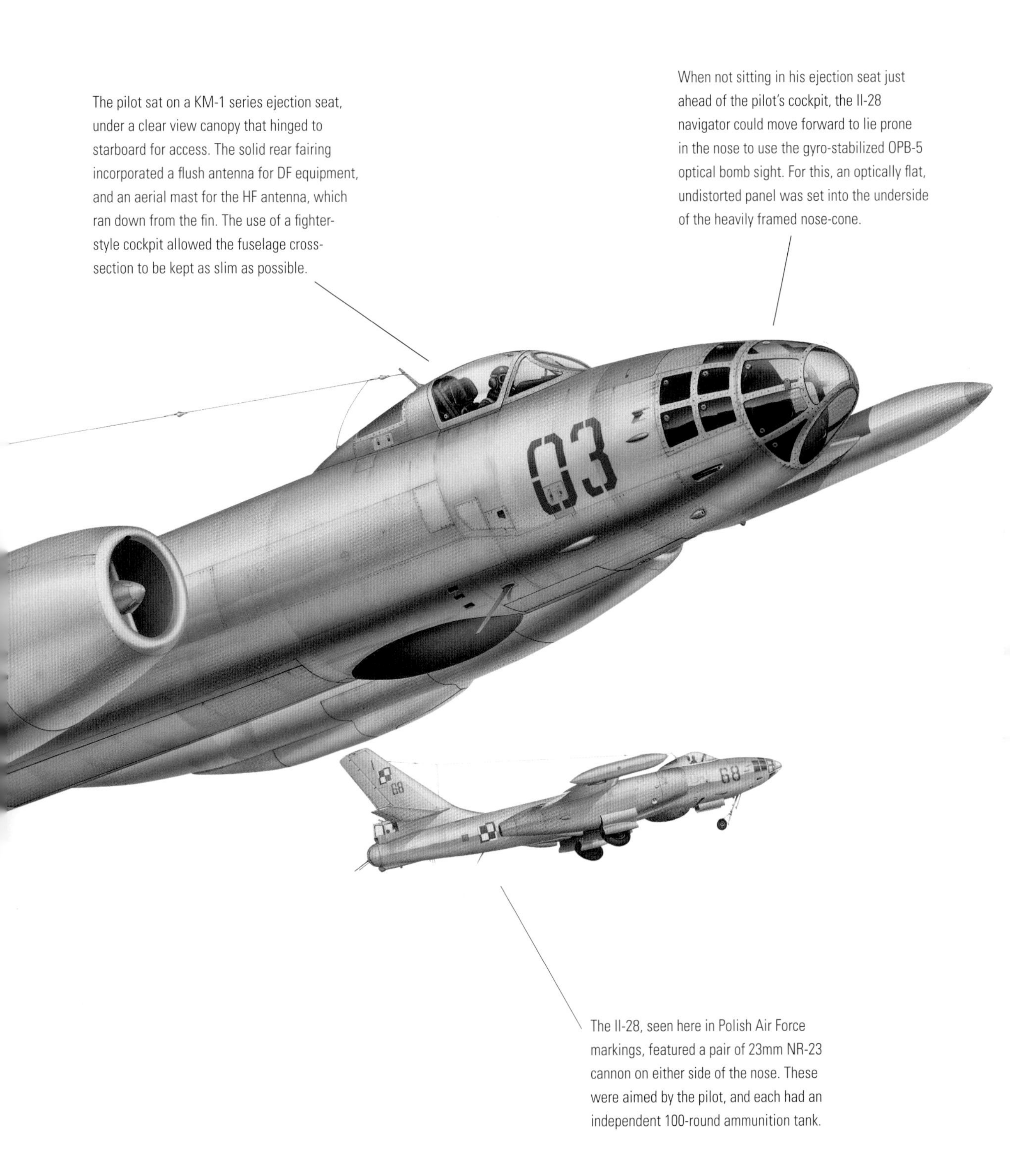

The Il-28, seen here in Polish Air Force markings, featured a pair of 23mm NR-23 cannon on either side of the nose. These were aimed by the pilot, and each had an independent 100-round ammunition tank.

Designed as a tactical light bomber to replace the piston-engined Tupolev Tu-2, Ilyushin's Il-28 formed the mainstay of the Soviet Bloc's tactical striking forces during the 1950s and was widely exported to countries within the Soviet sphere of influence. The first prototype Il-28 was fitted with RD-10 (Jumo 004) turbojets, but these failed to provide sufficient power and subsequent aircraft used the Klimov VK-1, a Russian copy of the Rolls-Royce Nene. The first VK-1-powered Il-28 flew on 20 September 1948 and deliveries to Soviet tactical squadrons began in the following year. Around 10,000 Il-28s were produced, variants including the Soviet

Opposite: **An Ilyushin Il-28 Beagle of the Chinese People's Air Force. Some 500 Il-28s were supplied to China, where the type was also built under licence as the Harbin H-5. The Il-28 began to be deployed to China during the Korean War, but took no part in the conflict.**

Navy's Il-28T torpedo bomber and the Il-28U two-seat trainer (NATO code-name Mascot). Some 500 Il-28s were supplied to China, where the type was also built under licence as the Harbin H-5. Egypt acquired 60, of which 20 were destroyed at Luxor by French F-84F Thunderstreaks during the Suez crisis of 1956. Il-28s formed part of the package of combat aircraft and missiles delivered to Cuba in 1962, provoking the so-called 'missile crisis' and a swift American response. A small number, thought to be about ten aircraft, were supplied to the North Vietnamese Air Force in the early 1960s, and the USAF deployed Convair F-102 interceptors to counter this potential threat, which in fact never materialized.

In its Il-28R reconnaissance version, the Beagle housed up to four optical cameras in a heated compartment within the redesigned bomb bay. Provision was also made for the aircraft to carry illumination flares for night photography, and the forward-firing armament was reduced to just one 23mm (0.90in) cannon in the left forward fuselage. The Il-28R carried additional fuel in its modified weapons bay and this, coupled with its carriage of tip tanks as standard, led to this variant being fitted with large main wheel tyres to compensate for the extra weight on take-off.

Hungary was the first nation to take the Il-28 into combat, ironically against the Soviet Union. On 23 October 1956 an uprising against the communist government began in Hungary, and Soviet forces moved in to crush the insurgents. The Hungarian Air Force initially played no part in the battle for independence but, following a 30 October declaration, combat operations against the Russians began the next day. The Il-28 took a relatively minor role, the Hungarians fielding a small force from their total of 40 Beagles in attacks against a pontoon bridge over the Tisza river. Soviet forces were back in control by 14 November. Egyptian Il-28s played a minor role in the various Arab–Israeli wars of the 1960s, more often than not as targets for Israeli attack aircraft. In the Nigerian civil war of 1967–70, six Il-28s, obtained from Soviet and Egyptian sources, were flown by Czech and Egyptian crews against Biafra on behalf of the Federal government. The Il-28s carried out an indiscriminate and destructive bombing campaign, although two and possibly three were destroyed in Biafran air attacks.

Left: **From its introduction into service the Il-28 was a successful and popular aircraft, thanks in no small measure to its Klimov VK-1 engines, which were copies of the Rolls-Royce Nene and extremely reliable. The photograph depicts three trials aircraft.**

Specification: Ilyushin IL-28	
Type:	three-crew tactical bomber
Powerplant:	two 2700kg (5953lb) thrust Klimov VK-1 turbojets
Performance:	maximum speed 902km/h (560mph) at sea level; service ceiling 12,300m (40,355ft); range 2180km (1355 miles)
Weights:	empty 12,890kg (28,422lb); maximum take-off 21,200kg (46,746lb)
Dimensions:	wing span 21.45m (70ft 4in); length 17.65m (57ft 11in); height 6.70m (22ft 0in); wing area 60.80m² (654.47 sq ft)
Armament:	four 23mm (0.90in) NR-23 cannon, two in nose and two in tail positions; up to 3000kg (6614lb) of bombs; two 400mm (15.7in) light torpedoes

Mikoyan-Gurevich MiG-17 Fresco

Located on the port side of the nose is a pair of Nudelmann-Richter NR-23 cannon. Eighty rounds of ammunition are carried for each gun. On the opposite side of the aircraft is a single 37mm cannon with 40 rounds.

Based at Maputo, Fresco-A 'Red 21' carries the markings of the Força Popular Aerea de Liberatação de Moçambique. The 24 MiG-17s in service before 1980 were joined by 12 delivered during 1983 and a further 12 delivered in March 1984. At least one was written off before 1980 and others were shot down by Renamo guerrillas, including two on 16 April 1985 and two on 6 October 1985.

The main fuselage fuel tank is located immediately aft of the cockpit. Further small tanks are located in the rear fuselage, either side of the jet-pipe. The MiG-17's endurance was poor, hence the carriage of wing tanks as standard fittings.

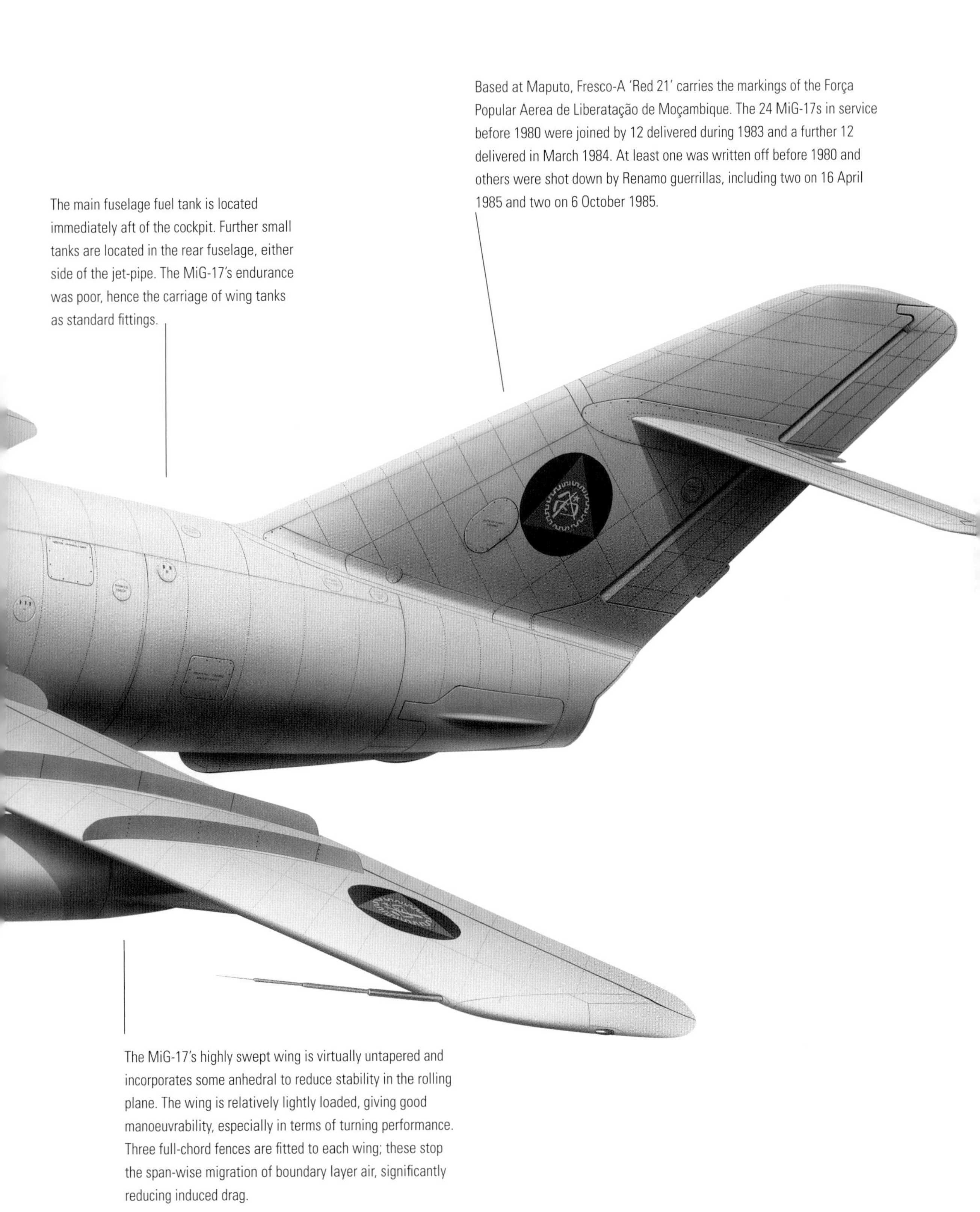

The MiG-17's highly swept wing is virtually untapered and incorporates some anhedral to reduce stability in the rolling plane. The wing is relatively lightly loaded, giving good manoeuvrability, especially in terms of turning performance. Three full-chord fences are fitted to each wing; these stop the span-wise migration of boundary layer air, significantly reducing induced drag.

When the MiG-17 first appeared in the early 1950s, Western observers at first believed that it was an improved MiG-15, with new features that reflected the technical lessons learned during the Korean War. In fact, design of the MiG-17 had begun in 1949, the new type incorporating a number of aerodynamic refinements that included a new tail on a longer fuselage and a thinner wing with different section and planform, with three boundary layer fences to improve handling at high speed.

The basic version, known to NATO as Fresco-A, entered service in 1952; this was followed by the MiG-17P all-weather interceptor (Fresco-B) and then the major production variant, the MiG-17F (Fresco-C), which had structural refinements and was fitted with an afterburner. The last variant, the MiG-17PFU, was armed with air-to-air missiles. Full-scale production of the MiG-17 in the Soviet Union lasted only five years before the type was superseded by the supersonic MiG-19 and MiG-21, but it has been estimated that around 8800 were built in that time, of which some 5000 were MiG-17Fs. It was the most numerous type in Frontal Aviation's tactical air defence units well into the 1960s, and remained in service until the late 1970s with training and reserve units. Even in 1980 it was probably true to say that nearly every Soviet fighter pilot had served some time on a MiG-17 unit. As the MiG-17 was replaced in Soviet service in the 1960s, many were released for export; it was in fact the most advanced type available from the Soviet Union throughout the 1960s, and until 1973/4, when multi-role versions of the MiG-21 and Su-7 were delivered outside the Warsaw Pact, the MiG-17 was the only Soviet fighter-bomber available for export, the early MiG-21s being pure interceptors. MiG-17s saw action in the Congo, in the Nigerian civil war and in the Middle East, where they were the workhorse fighter-bombers of the Syrian Air Force. Some of these aircraft were fitted with six underwing hardpoints, as on the licence-built Polish version, the LIM-5M, mounting inboard pylons for bombs, drop-tanks in the usual locations, and rails for unguided rocket projectiles.

Below: **The MiG-17, seen here in fighter-bomber configuration, was a serious problem over North Vietnam for more modern US attack aircraft such as the F-4 Phantom, which could be out-manoeuvred by the Russian-built fighter and had to rely on superior speed to disengage.**

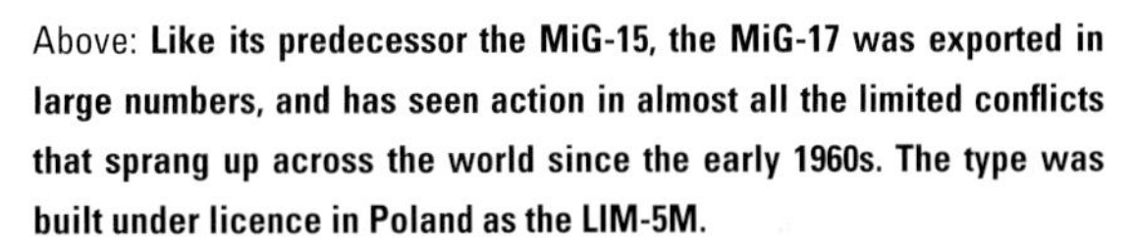
Above: **Like its predecessor the MiG-15, the MiG-17 was exported in large numbers, and has seen action in almost all the limited conflicts that sprang up across the world since the early 1960s. The type was built under licence in Poland as the LIM-5M.**

The MiG-17's claim to historical importance rests on its use by North Vietnamese Air Force (NVAF) units between 1965 and 1973. When the USAF began to plan a bombing campaign against North Vietnam in 1964, the NVAF had no fighter force; this situation began to change in late 1964, with the arrival of MiG-15s and J-5s (licence-built MiG-17s) from China. Soviet-supplied MiG-17s arrived in the following year, and along with further J-5s these were the most important. NVAF fighters fought throughout most of the war. The MiG-15 was little used, while the MiG-21 was not available in large quantities. It came as a severe shock to the USAF and US Navy to discover that the MiG-17 was a serious match for their modern, radar-equipped, automated supersonic fighters. The MIG-17 had a far lower wing loading than its opponents; its span was almost as great as that of the F-4 Phantom, but its loaded weight was about the same as that of the F-4's internal fuel load. The result was that the MiG-17 could enter a relatively slow but very tight turn that no US aircraft could match.

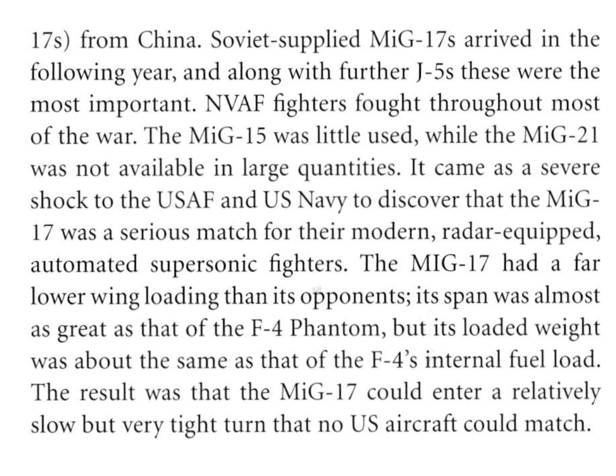

Specification: MiG-17F Fresco-C	
Type:	single-seat fighter
Powerplant:	one 3383kg (7459lb) thrust Klimov VK-1F turbojet
Performance:	maximum speed 1145km/h (711mph) at 3000m (9840ft); service ceiling 16,600m (54,461m); range 1470km (913 miles) with slipper tanks
Weights:	empty 4100kg (9040lb); maximum take-off 6000kg (13,230lb)
Dimensions:	wing span 9.45m (31ft); length 11.05m (36ft 3in); height 3.35m (11ft); wing area 20.60m^2 (221.74 sq ft)
Armament	one 37mm (1.45in) N-37 and two 23mm (0.90in) NS-23 cannon; up to 500kg (1102lb) of underwing stores

Sukhoi Su-7 Fitter

In Russian service the Su-7 soldiered on into the 1990s in a range of test roles. Specially modified aircraft included this Su-7LL ejection seat test bed, which was based on an Su-7UM airframe with the rear cockpit modified for the installation of test articles.

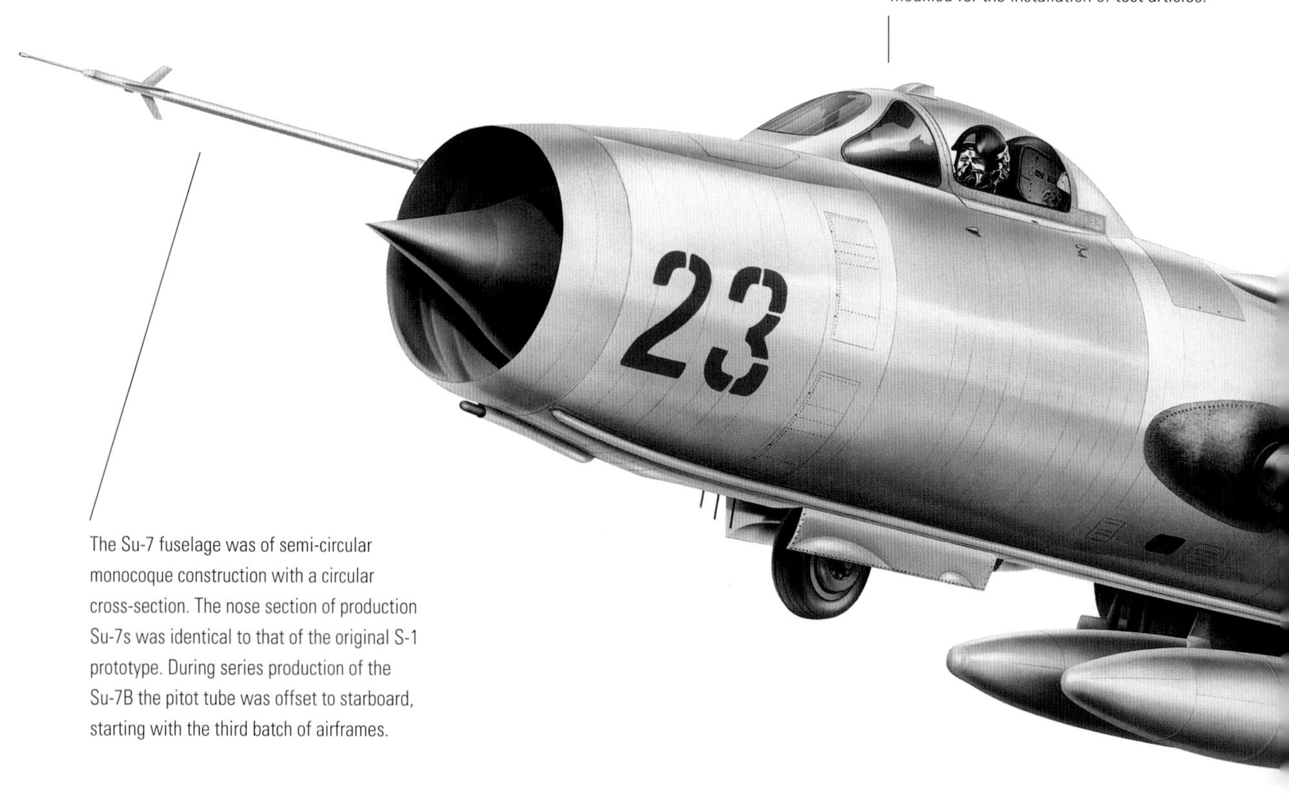

The Su-7 fuselage was of semi-circular monocoque construction with a circular cross-section. The nose section of production Su-7s was identical to that of the original S-1 prototype. During series production of the Su-7B the pitot tube was offset to starboard, starting with the third batch of airframes.

Production Su-7s were armed with a pair of 30mm (1.18in) NR-30 cannon. One weapon within each wing root leading edge was provided with 70 rounds of ammunition. The same twin-gun armament was carried by all production variants, including the early two-seat Su-7U and Su-7UM.

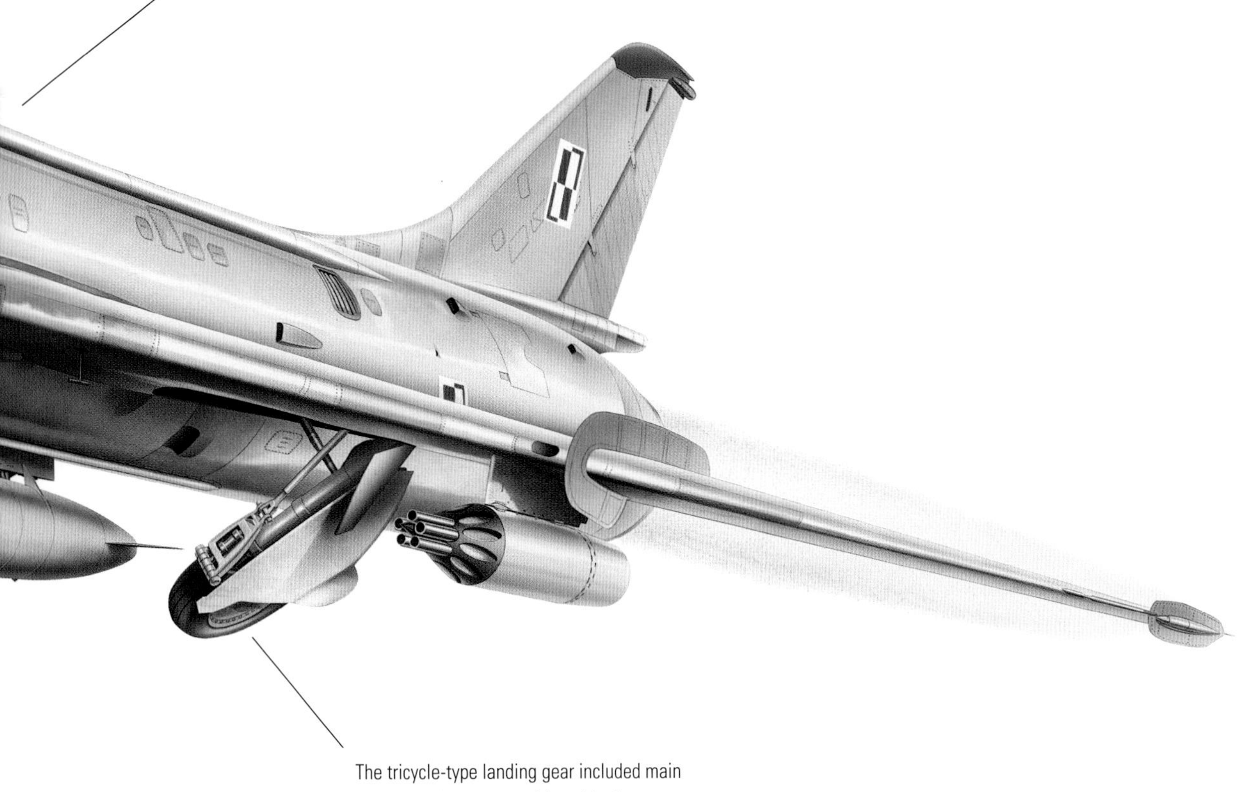

The tricycle-type landing gear included main gear units that were positioned in the outer wings and equipped with levered suspension wheels with oleo-pneumatic shock absorbers. The Su-7BM was fitted with new KT-69 braked wheels.

Above: **The Su-17M in its latter variants formed an important part of the tactical elements of the Soviet Air Force prior to the collapse of the USSR. The nearest aircraft carries a typical ground attack load of a single S-24 rocket on the starboard inboard wing station and a pair of FAB-250 bombs under the fuselage.**

Pavel Sukhoi, who had earlier fallen from favour with Stalin and been reinstated after the latter's death in 1953, contributed greatly to the Russian air build-up in the late 1950s. From his design bureau came the Su-7 Fitter-A and the Su-9 Fishpot-A, both single-engined, single-seat types. The Su-7, first seen in public in 1956 and designed for close air support with the Frontal Aviation, featured 60 degrees of sweep and carried two 30mm (1.18in) cannon in its wing roots; it also carried a relatively heavy load of ordnance, either rockets or bombs, under the wings. The Su-7 remained the Soviet Air Force's standard tactical fighter-bomber throughout the 1960s. Later, the Sukhoi bureau redesigned the Su-7, giving it a more powerful engine, variable-geometry wings and increased fuel tankage. In this guise it became the Su-17/20 Fitter C, which was unique among combat aircraft in being a variable-geometry derivative of a fixed-wing machine. It was an excellent example of a remarkable Russian talent for developing existing designs to their fullest extent. The development of the Fitter-C was a facet of the Soviet practice of constant development, enabling them to keep one basic design of combat aircraft in service for 30 or 40 years and foster long-term standardization. Also, the use of the same production facilities over a long period of time helped greatly to reduce costs, which is why the USSR was able to offer combat types on the international market at far more competitive rates than the West. The Su-22 was an updated version with terrain-avoidance radar and other improved avionics, while the Su-7U, code-named Moujik, was a two-seat trainer version.

It has been revealed that pilots of Su-7s in service with the Polish Air Force were trained for the tactical nuclear role. A group of pilots and technicians from 5 Regiment 'Pomorski' received tactical nuclear weapons training in the USSR. 'Special attack' procedure with practice bombs was undertaken by Polish Su-7s during the Warsaw Pact Exercise October Storm, near Erfurt, East Germany, in 1965. Polish Su-7 operations were conducted under Soviet Frontal Aviation control, with the RU-57 tactical nuclear bomb being carried on one of the fuselage weapon stations, plus three external fuel tanks (one on the remaining under-fuselage hardpoint and one under each wing). As such, the Polish Su-7s effectively assumed a clandestine nuclear role on behalf of the USSR within the framework of the Polish Air Force. Polish AF Su-7s were based close to Torun, where tactical nuclear weapons were stored in Soviet bunkers. Polish units equipped with the Su-7BMK and Su-7UMK were stationed at Bydgoszcz until shortly before the base was closed in 1969.

The Su-7 saw a good deal of action with the Egyptian Air Force, initially during the Six-Day War of 1967, when two aircraft were shot down during a bombing raid on El Arish. In the so-called War of Attrition of 1969–70, an Egyptian Su-7 belly-landed intact near the Ghidi Pass in November 1969, providing Israel with a valuable intelligence coup. Three Su-7s were lost during a hit-and-run raid across the Suez Canal in April 1970, and on 11 September 1970 one example was lost during a reconnaissance sortie. Eight squadrons of Egyptian AF Su-7BMs (backed up by three Algerian units) were involved in the Yom Kippur War of 1973, supporting the 900-tank assault on the Golan Heights on 6 October, and strafing Israeli columns in concert with MiG-17s and Iraqi Hunters during the opening phase. The Su-7 was credited with a good combat record during the conflict, proving remarkably resistant to ground fire.

Above: **Poland's Air Defence and Aviation Force operates 80 Su-22M-4s and 15 Su-22M-3s, delivered from 1985. The aircraft are divided between the 8th Air Brigade at Miroslaviec and the 39th and 40th Air Brigades at Swidin Smardzko, and the 6th and 7th Tactical Aviation Squadrons at Powidz.**

Contemporary with the Su-7, the Su-9 Fishpot-A was a single-seat interceptor – to some extent an Su-7 with a delta wing. It was armed with the first Soviet AAM, the semi-active radar-homing Alkali, four of which were carried under the wings. In 1961 a new model, the Su-11 Fishpot-B, was developed from the Su-9, and was followed into service by the Fishpot-C, which had an uprated engine.

Specification: Sukhoi Su-7B Fitter-A

Type:	single-seat tactical fighter-bomber
Powerplant:	one 9008kg (19,862lb) thrust Lyulka AL-7F turbojet
Performance:	maximum speed 1700km/h (1055mph) at altitude; service ceiling 15,200m (49,865ft)
Combat radius:	320km (199 miles) hi-lo-hi with 50 per cent load
Weights:	empty 8620kg (19,000lb); maximum take-off 13,500kg (29,750lb)
Dimensions:	wing span 8.93m (29ft 3in); length 18.75m (61ft 6in); height 5.00m (16ft 5in); wing area 27.60m2 (297.09 sq ft)
Armament:	two 30mm (1.18in) NR-30 cannon; four external pylons with provision for two 750kg (1653lb) and two 500kg (1102lb) bombs

Mikoyan-Gurevich MiG-21 Fishbed

Twin forward airbrakes open out and down diagonally, under the thrust of hydraulic rams, to slow the aircraft. A third airbrake is mounted aft. Opening the airbrakes causes no change in aircraft trim. The rear airbrake, on the centreline, is bulged and perforated; like the two forward ones, it opens against the airflow under the force of a hydraulic jack.

Formerly belonging to No 7 Squadron (Battle Axes) of the Indian Air Force, this MiG-21MF was one unit in India's significant MiG force. Most of the IAF MiG-21 units subsequently converted to the Dassault Mirage 2000.

The conical centrebody of the engine inlet is mounted on rails and can be slid in and out hydraulically to three positions: retracted (normal), part extended (Mach 1.5 plus) and fully extended (Mach 1.9 plus). Inside is the R2L 'Jay Bird' radar suite.

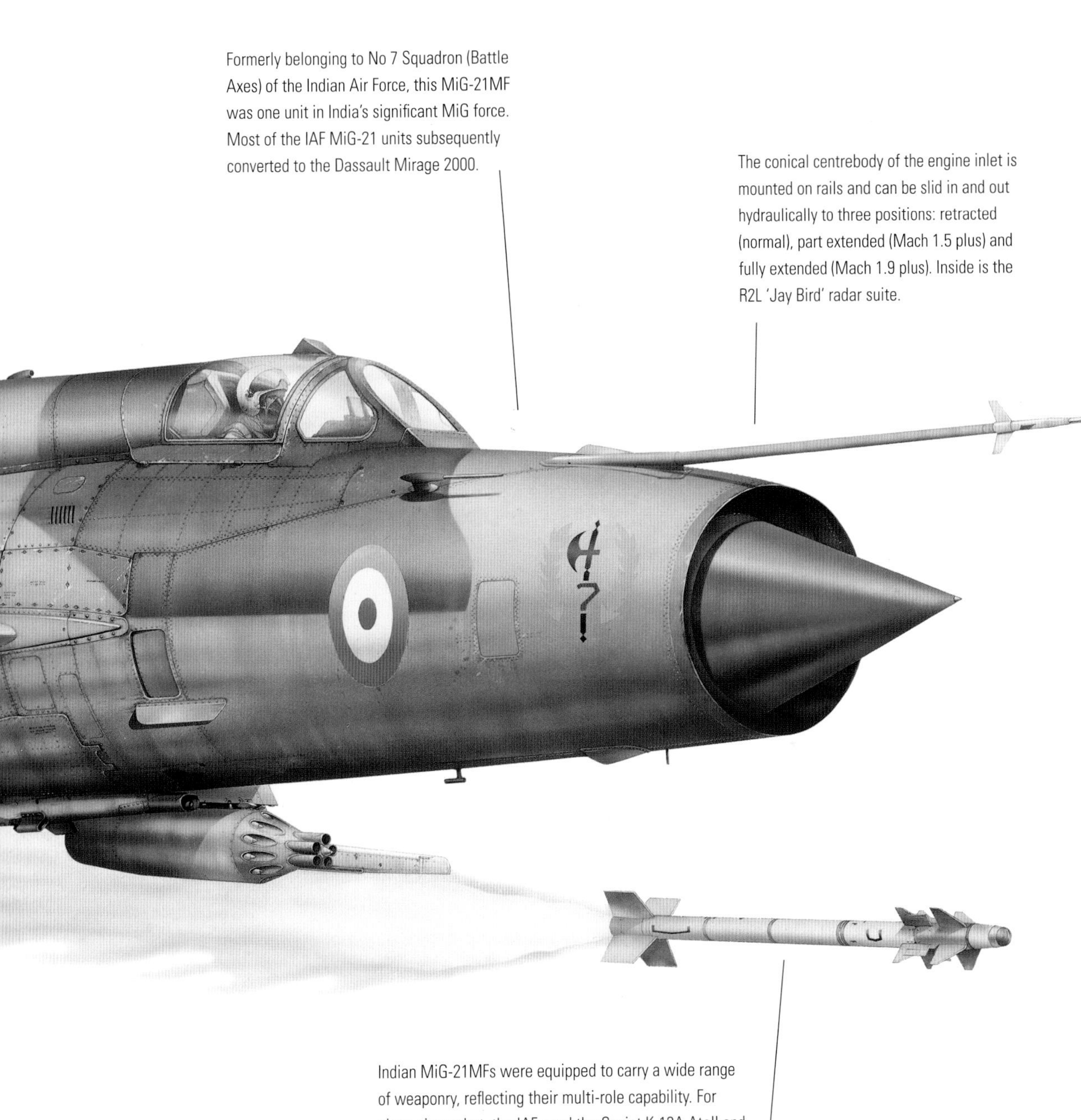

Indian MiG-21MFs were equipped to carry a wide range of weaponry, reflecting their multi-role capability. For air-to-air combat, the IAF used the Soviet K-13A Atoll and R-60 Aphid, as well as the French Matra R550 Magic, seen here. Most MiG-21 fighters were armed with the GSh-23L cannon, with a pair of 23mm calibre barrels, recessed into the bottom of the fuselage.

The MiG-21 was a child of the Korean War, where Soviet air combat experience had identified a need for a light, single-seat target defence interceptor with high supersonic manoeuvrability. Two prototypes were ordered, both appearing early in 1956; one, code-named Faceplate, featured sharply swept wings and was not developed further. The initial production versions (Fishbed-A and -B) were built only in limited numbers, being short-range day fighters with a comparatively light armament of two 30mm (1.18in) NR-30 cannon, but the next variant, the MiG-21F Fishbed-C, carried two K-13 Atoll infrared homing AAMs, and had an uprated Tumansky R-11 turbojet as well as improved avionics. The MiG-21F was the first major production version; it entered service in 1960 and was progressively modified and updated over the years that followed. In the early 1970s the MiG-21 was virtually redesigned, re-emerging as the MiG-21B (Fishbed-L) multi-role air superiority fighter and ground attack version. The Fishbed-N, which appeared in 1971, introduced new advanced construction techniques, greater fuel capacity and updated avionics for multi-role air combat and ground attack. In its several versions the MiG-21 became the most widely used jet fighter in the world, being licence-built in India, Czechoslovakia and China, where it was designated the Shenyang F-8, and equipping some 25 Soviet-aligned air forces. A two-seat version, the MiG-21U, was given the NATO reporting name Mongol.

In Vietnam the MiG-21 was the Americans' deadliest opponent. The tactics employed by the MiG pilots to intercept US aircraft operating over the North involved flying low and then zooming up to attack the heavily laden fighter-bombers, mainly F-105 Thunderchiefs, forcing them to jettison their bomb loads as a matter of survival. To counter this, Phantoms armed with Sidewinder AAMs flew at lower altitudes than the F-105s, enabling the crew to sight the MiGs at an early stage in their interception attempt and then use the Phantom's superior speed and acceleration to engage the enemy. These were very much in the nature of hit-and-run tactics, the Phantom pilots avoiding turning combat because of the MiG-21's superior turning ability, but they worked; the Phantom crews had a superb early warning facility in the shape of EC-121 electronic surveillance aircraft, which were able to direct the MiGCAP fighters on to their targets in good time. In 1966 American fighters destroyed 23 MiGs for the loss of nine of their own aircraft.

Neither the MiG-17 nor the MiG-21 was a specialized night fighter, but they often operated in that capacity, particularly when the American night bombing campaign against the North intensified during 1972. One US Navy pilot, Captain R.E. Tucker, recalled that: 'During 1972, when Navy A-6s were making a number of single aircraft low-level night strikes all over the area between Haiphong and Hanoi, MiGs were known to launch, making the A-6 drivers nervous (although I personally felt that the MiGs had no night/IFR capability against an A-6 at 300 feet). As a result, we started positioning a single F-4 on MiGCAP along the coastline at night. We figured a single F-4 didn't have to worry about a mid-air with his wingman and had plenty of potential at night against a single MiG. If a MiG launched and headed towards an A-6, the F-4 would vector for the MiG. Invariably MiGs would run for home when the F-4 got to within 25–30 miles of them. Some pilots weren't too overjoyed about the night MiGCAP missions, but I personally felt it was a golden opportunity and my MiG kill proved it. I figured that the MiG had a negligible opportunity to do anyone bodily harm at night with his limited weapons system and Atoll/guns load, whereas the

Above: **The MiG-21MF Fishbed-J was a derivative of the MiG-21M for exclusive service with the Soviet Air Force, and introduced air-to-air capability on all four underwing points. Even in the early 21st century, the MiG-21 remains an opponent to be reckoned with.**

Below: **This MiG-21M is seen in the colours of the Romanian Air Force, which operates two groups of the type: Grupul 95 based at Bacau, and Grupul 86, based at Fetesti. The RomAF's MiG-21Ms were all built after 1975, and represent the most modern 'new-build' aircraft of this version.**

Left: **The second-generation MiG-21s, such as those seen here in Czechoslovak Air Force service, were endowed with progressively heavier armament and increasingly sophisticated avionics. All had offset pitot probes, blown flaps, two-piece canopies and broad-chord tail fins.**

F-4 had a good solid head-on or tail shot against any MiG he could find and get close enough to shoot at.' Tucker, then a Lieutenant-Commander, was flying an F-4 Phantom of VF-103, operating from USS Saratoga, when he destroyed a MiG-21 on the night of 10/11 August 1972. The Phantom was armed with two AIM-7E Sparrows and two AIM-9D Sidewinders. Guided towards the target by his RIO, Lt (jg) Bruce Edens, at a range of about two miles the pilot launched two Sparrows, the second one leaving the under-fuselage rack just as the first warhead detonated. 'There was a large fireball', Tucker reported, 'and the second missile impacted in the same spot. I came right slightly to avoid any debris. The target on our radar appeared to stop in mid-air and within a second or two the radar broke lock. The MiG-21 pilot did not survive. If he ejected after the first missile, the second missile must have done him in. We couldn't see any debris in the dark ... The kill was confirmed about three days later.'

The MiG-21 deserves its place in history as one of the most versatile combat aircraft of the post-World War II era. In designing the type, Mikoyan hit upon the lucky formula of combining a delta wing with swept tail surfaces, a combination that made the aircraft light and very manoeuvrable.

Specification: MiG-21MF Fishbed-J	
Type:	single-seat multi-role combat aircraft
Powerplant:	one 7500kg (16,535lb) thrust Tumanskii R-13-300 turbojet
Performance:	maximum speed 2229km/h (1385mph) at 11,000m (36,090ft); service ceiling 17,500m (57,400ft); range 1160km (721 miles) on internal fuel
Weights:	empty 5200kg (11,466lb); maximum take-off 10,400kg (22,932lb)
Dimensions:	wing span 7.15m (23ft 5in); length 15.76m (51ft 8in); height 4.10m (13ft 5in); wing area 23.0m² (247.58 sq ft)
Armament:	one 23mm (0.90in) GSh-23L twin-barrel cannon in pack under fuselage; four underwing pylons with provision for 1500kg (3307lb) of stores, including AAMs, rocket pods, napalm tanks and drop-tanks

Mikoyan MiG-25 Foxbat

Each of these wingtip pods is a tube almost 30cm (12in) in diameter. Part of its length is filled with heavy metal to serve as a mass, which damps out wing flutter at speed; the rest houses avionics. Sirena 3 warning receivers are fitted in each pod, looking outwards to cover the sector to left and right.

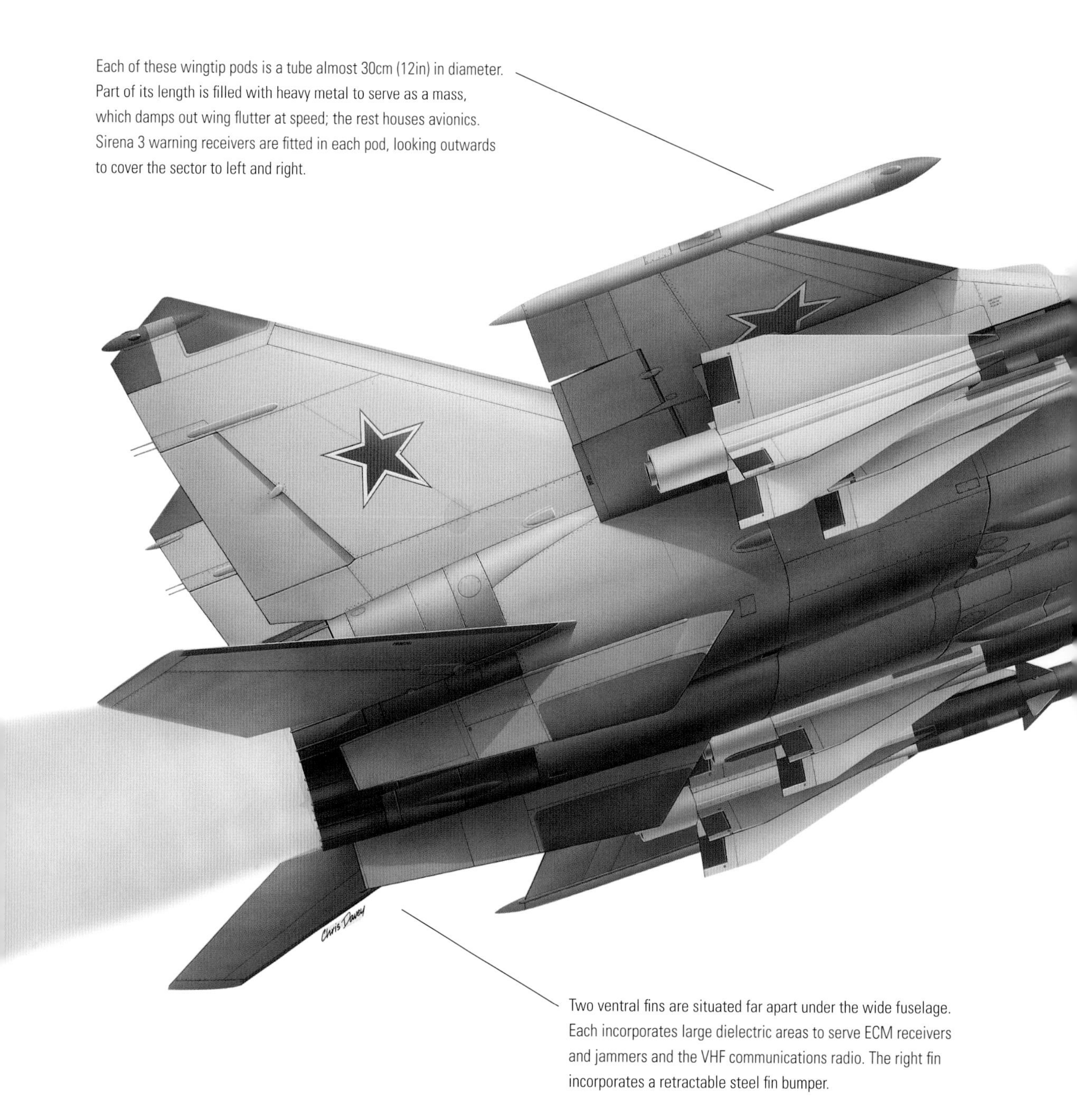

Two ventral fins are situated far apart under the wide fuselage. Each incorporates large dielectric areas to serve ECM receivers and jammers and the VHF communications radio. The right fin incorporates a retractable steel fin bumper.

The standard air-to-air missiles carried by the MiG-25 are of the AA-6 'Acrid' family. The biggest AAM in the world, the AA-6 comes in two versions, infrared and semi-active radar homing (SARH). This aircraft has SARH missiles on the outboard pylons, which steer towards radar signals reflected back from the target, and round-nose IR-homing missiles on the inner pylons.

The MiG-25P's enormous main radar, codenamed Fox Fire by NATO, is a typical 1959 technology set. It uses thermionic valves (vacuum tubes) and puts out 600kW of power to burn through enemy jamming. The tip of the radome has a steel tube that carries a pitot/static system to dissipate static electricity, the SP-50 'Swift Rod' ILS aerial and pitch/yaw transducers feeding the air data system.

Above: **The no-frills nature of the MiG-25 Foxbat is apparent in this photograph of an aircraft, which has become a museum piece. The MiG-25 was developed in great haste to counter a new generation of American supersonic bombers, which in fact never materialized.**

The prototype MiG-25 was flown as early as 1964 and was apparently designed to counter the projected North American B-70 bomber, with its Mach 3.0 speed and ceiling of 21,350m (70,000ft). The cancellation of the B-70 left the Foxbat in search of a role; it entered service as an interceptor in 1970 with the designation MiG-25P (Foxbat-A), its role now redefined as being capable of countering all air targets in all weather conditions, by day or by night, and in dense hostile electronic warfare environments. The MiG-25P continues to serve in substantial numbers, and constitutes part of the Russian S-155P missile interceptor system. The aircraft is produced by RAC MiG (formerly MAPO-MiG), which is based in Moscow, and the Sokol Aircraft Manufacturing Plant Joint Stock Company, at Nizhni Novgorod in Russia. Variants of the MiG-25 are also in service in Ukraine, Kazakhstan, Azerbaijan, India, Iraq, Algeria, Syria and Libya.

The MiG-25P is a twin-finned, high-wing monoplane with slightly swept wings and a variable-incidence tailplane. To improve the aircraft's longitudinal stability, and to avert stall at steep angles and subsonic speed, there are two shallow upper-surface fences on each wing. The high-wing monoplane configuration, together with lateral air intakes, has the effect of reducing the loss of aerodynamic efficiency resulting from wing-fuselage interference.

The aircraft is armed with four R-40 (NATO codename AA-6 Acrid) air-to-air missiles fitted with infrared and radar homing heads. The missiles are suspended from four underwing pylons. The MiG-25P may also be fitted with two R-40 and four R-60 (AA-8 Aphid), or two R-23 (AA-7 Apex) and four R-73 (AA-11 Archer). The MiG-25 is not fitted with a gun. Electronic equipment includes the Smerch-A2 radar sight (NATO codename Fox Fire), developed by

Above: **The basic MiG-25 has been greatly refined since it first entered service as an interceptor in the 1970s. One of its most important roles is that of reconnaissance, the aircraft's high speed and altitude putting it out of reach of many defensive systems.**

the Phazotron Research and Production Company; Identification Friend/Foe (IFF) transponder; aircraft responder to maintain communications with guidance and landing radars operating in active radio-location mode; and radar warning receiver. The flight control and navigation equipment includes the ARK-10 automatic radio compass, RV-4 radio altimeter and Polyot-11 navigation and landing system. This system, coupled with ground radio beacon and landing radio beacons, provides programmed aircraft manoeuvres including climb, en-route flight, returns to the departure airfield or to one of three selected emergency airfields, and low-run landing approach and missed approach manoeuvres.

The MiG-25 is powered by two Tumanskii R-15B-300 single-shaft turbojets, arranged in the tail section of the fuselage. Fuel is supplied by two welded structural tanks situated between the cockpit and engine bay and occupying about 70 per cent of the fuselage volume, in saddle tanks around the engine ducts, and in an integral tank in each wing, filling almost the entire volume inboard of the outer fence. During the 1991 Gulf War, a MiG-25 was the only Iraqi aircraft to score an aerial victory, shooting down an F/A-18 Hornet. On the few occasions when MiG-25s were encountered, they proved capable of out-running both the F-15 Eagle and the latter's AIM-7 air-to-air missiles.

The MiG-25R, MiG-25RB and MiG-25BM are derivatives of the MiG-25P. The MiG-25R, as its suffix implies, is a reconnaissance variant, while the MiG-25RB has a high-level bombing capability against area targets. This version is fitted with a reconnaissance station, aerial camera, topographic aerial camera, the Peteng sighting and navigation system for bombing programmed targets, and electronic countermeasures (ECM) equipment, which includes active jamming and electronic reconnaissance systems. The MiG-25BM variant has the capability to launch guided missiles against ground targets, and to destroy area targets, targets with known coordinates, and enemy radars. Its principal anti-radar weapon is the Kh-58 (AS-11 Kilter), developed and manufactured by the Raduga Engineering Design Bureau, Moscow.

The MiG-25 interceptor is gradually being replaced in first-line service by a greatly improved version, the MiG-31. First flown on 16 September 1975 as the Ye-155MP, and originally designated MiG-25MP, the MiG-31 (known by the NATO reporting name Foxhound) entered production in 1975 and the first units to arm with the type became operational in 1982, replacing MiG-23s and Su-15s.

Specification: MiG-25P Foxbat-A	
Type:	single-seat interceptor
Powerplant:	two 10,200kg (22,491lb) thrust Tumanskii R-15B-300 turbojets
Performance:	maximum speed 2974km/h (1847mph) at altitude; service ceiling 24,383m (80,000ft); combat radius 1130km (702 miles)
Weights:	empty 20,000kg (44,100lb); maximum take-off 37,425kg (82,520lb)
Dimensions:	wing span 14.02m (46ft 0in); length 23.82m (78ft 1in); height 6.10m (20ft 0in); wing area 61.40m² (660.9 sq ft)
Armament:	four underwing pylons for various combinations of air-to-air missile

Sukhoi Su-25 Frogfoot

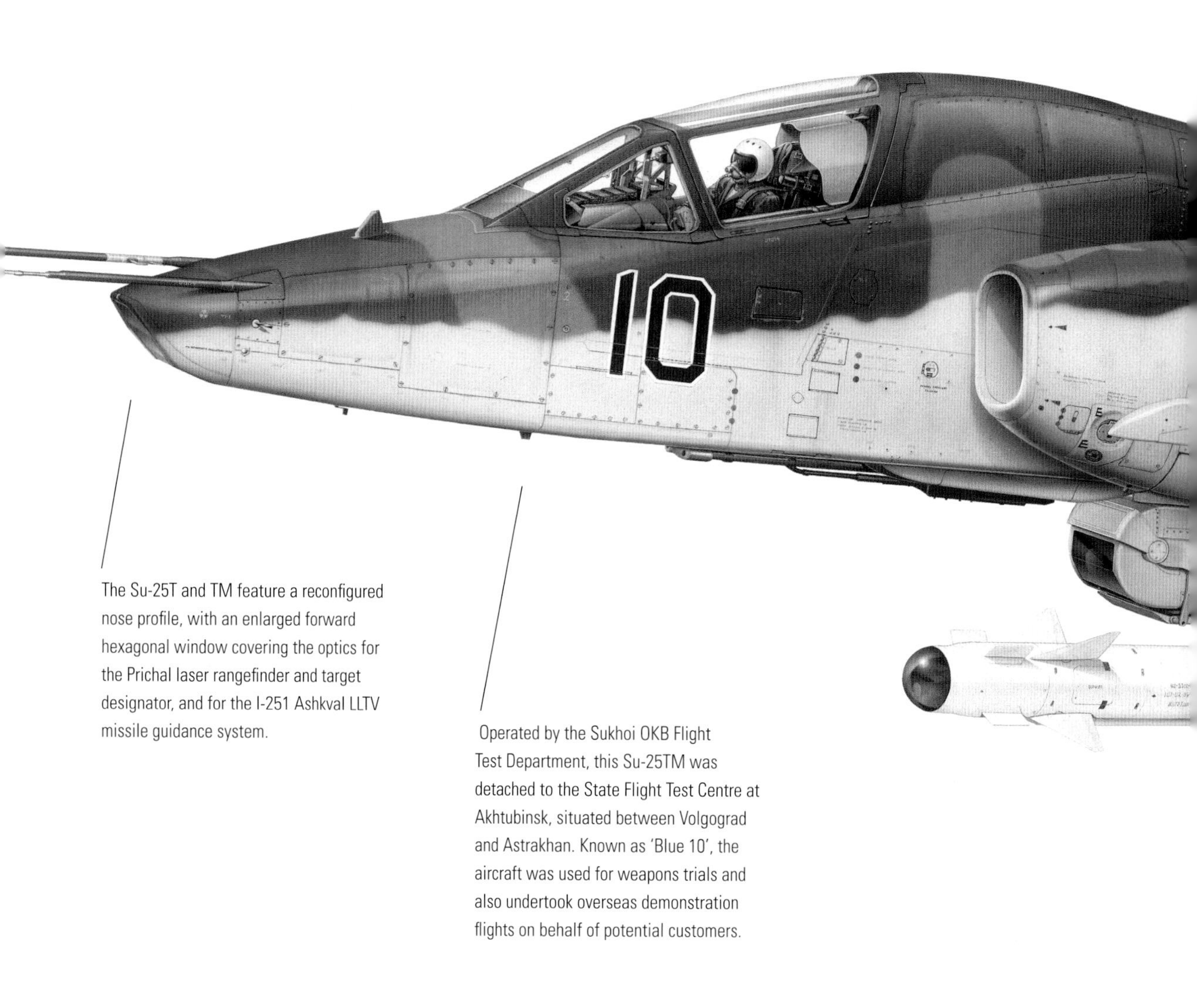

The Su-25T and TM feature a reconfigured nose profile, with an enlarged forward hexagonal window covering the optics for the Prichal laser rangefinder and target designator, and for the I-251 Ashkval LLTV missile guidance system.

Operated by the Sukhoi OKB Flight Test Department, this Su-25TM was detached to the State Flight Test Centre at Akhtubinsk, situated between Volgograd and Astrakhan. Known as 'Blue 10', the aircraft was used for weapons trials and also undertook overseas demonstration flights on behalf of potential customers.

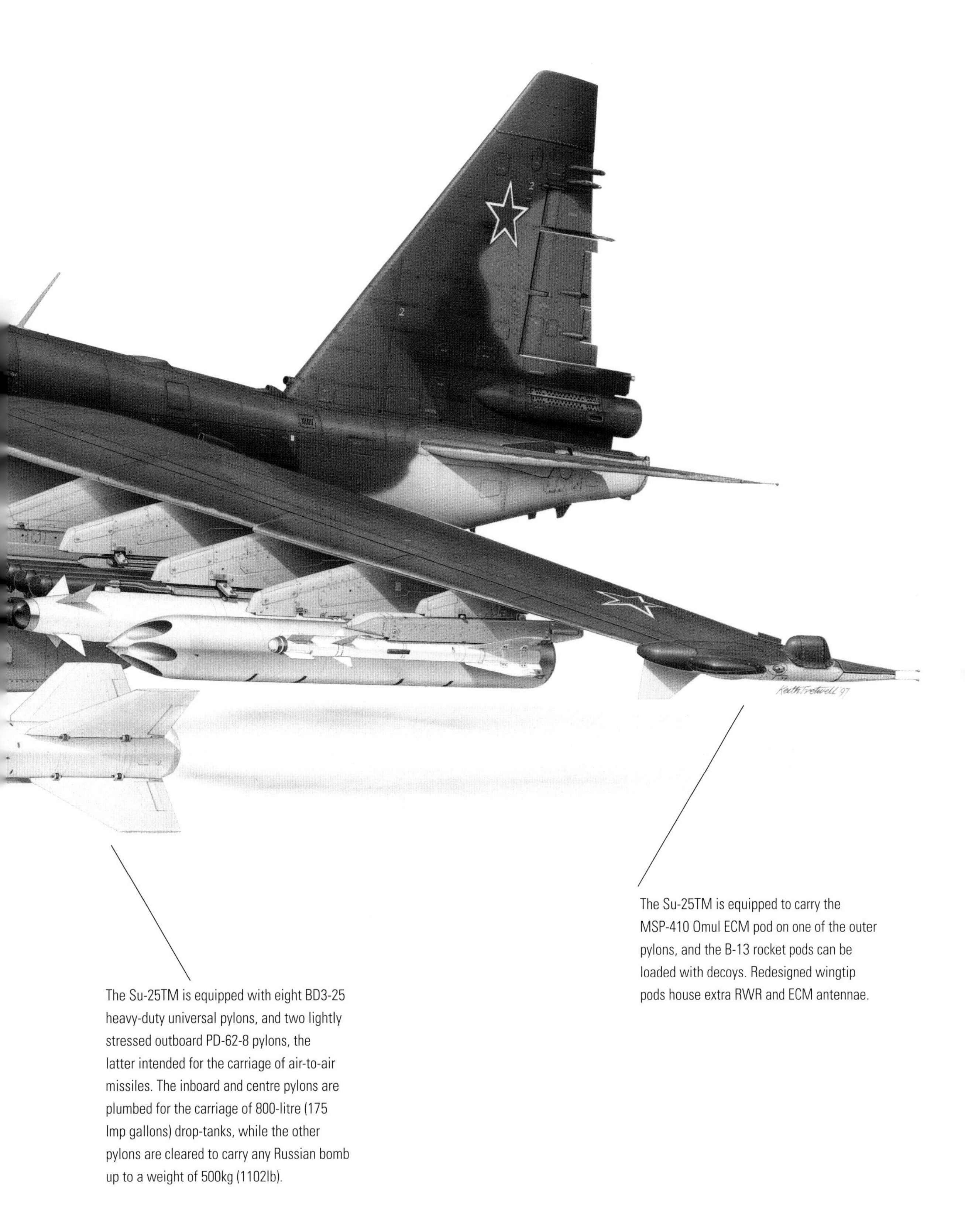

The Su-25TM is equipped to carry the MSP-410 Omul ECM pod on one of the outer pylons, and the B-13 rocket pods can be loaded with decoys. Redesigned wingtip pods house extra RWR and ECM antennae.

The Su-25TM is equipped with eight BD3-25 heavy-duty universal pylons, and two lightly stressed outboard PD-62-8 pylons, the latter intended for the carriage of air-to-air missiles. The inboard and centre pylons are plumbed for the carriage of 800-litre (175 Imp gallons) drop-tanks, while the other pylons are cleared to carry any Russian bomb up to a weight of 500kg (1102lb).

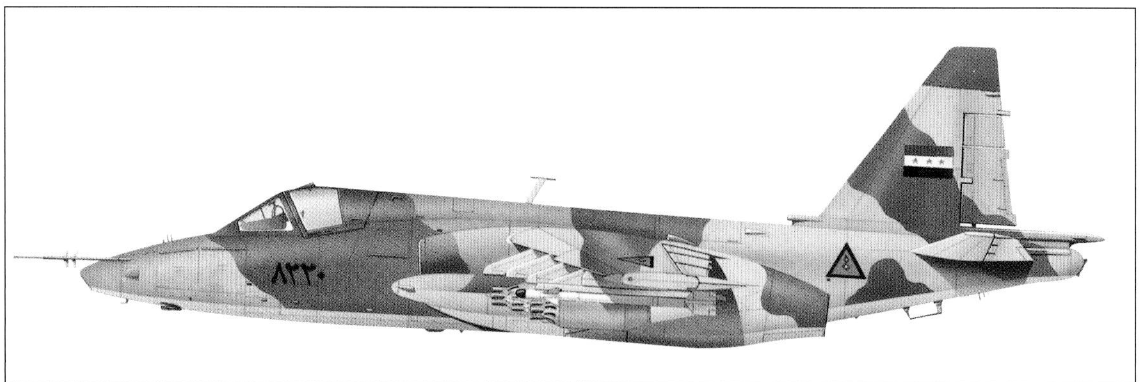

Above: **Delivered in 1986–87, Iraq's 30 to 45 'Frogfoots' carry a unique sand and earth camouflage, with pale blue undersides. First displayed in public in May 1989, the aircraft may have been used in the war with Iran, and two were shot down by a USAF F-15C during Operation Desert Storm.**

A Russian requirement for an attack aircraft in the A-10 Thunderbolt II class materialized in the Sukhoi Su-25, although in truth the design of the Su-25 – allocated the NATO reporting name Frogfoot – approximates more closely to the Dassault-Dornier Alpha Jet or the British Aerospace

Above: **With its wingtip airbrakes deployed, this Su-25TM from the Akhtubinsk Scientific and Technical Institute Flying Centre represents the latest combat variant of the Frogfoot series. The aircraft has improved defensive systems to counter man-portable SAMs.**

Hawk. Deployment of the single-seat close-support Su-25K began in 1978, and the aircraft saw considerable operational service during the former Soviet Union's involvement in Afghanistan and the ruggedness of the design was revealed in dramatic fashion on numerous occasions. One particular aircraft, flown by Colonel Alexander V. Rutskoi, was actually heavily damaged on two occasions, one by anti-aircraft fire and then by Sidewinder AAMs launched by Pakistani AF F-16s. On each occasion the pilot managed to limp back to base. The aircraft was repaired, repainted and returned to service. Rutskoi was less lucky; while flying a second Su-25 on a combat mission, his aircraft was hit by anti-aircraft fire and a Blowpipe shoulder-launched missile, which exploded in the starboard engine. The aircraft still flew, but another burst of AA brought it down. Rutskoi ejected and spent some time as a prisoner of the Pakistani authorities before being repatriated. However, operations in Afghanistan also revealed a number of serious shortcomings. For example, the close positioning of the Su-25's engines meant that if one took a hit and caught fire, the other was likely to catch fire too. When the Frogfoot first encountered the Stinger shoulder-launched missile, four aircraft were shot down in two days, with the loss of two pilots; it was established that missile fragments shredded the rear fuselage fuel tank, which was situated directly above the jet exhaust.

As a result of lessons learned during the Afghan conflict an upgraded version known as the Su-25T was produced, with improved defensive systems to counter weapons like the Stinger. The improvements included the insertion of steel plates, several millimetres thick, between the engine bays and below the fuel cell. After this modification no further Su-25s were lost to shoulder-launched missiles. In total, 22 Su-25s were lost in the nine years of the Afghan conflict.

The Su-25UBK is a two-seat export variant, while the Su-25UBT is a navalized version with a strengthened undercarriage and arrester gear. The Su-25UT (Su-28) was a trainer version, lacking the weapons pylons and combat capability of the standard Su-25UBK, but retaining the original rough field capability and endurance. It was planned as a replacement for the huge numbers of Aero L-29 and L-39 trainers in service with the former Soviet Air Force, but only one aircraft was flown in August 1985, appearing in the colours of DOSAAF, the Soviet Union's paramilitary 'private flying' organization, which provided students with basic flight training. The aircraft, which actually outperformed the L-39, appeared in many aerobatic displays.

Specification: Sukhoi Su-25 Frogfoot A

Type:	single-seat close support aircraft
Powerplant:	two 4500kg (9922lb) Tumanskii R-195 turbojets
Performance:	maximum speed 975km/h (605mph) at sea level; service ceiling 7000m (22,965ft); combat radius 750km (466 miles) lo-lo-lo with 4400kg (9700lb) war load
Weights:	empty 9500kg (20,950lb); maximum take-off 17,600kg (38,800lb)
Dimensions:	wing span 14.36m (47ft 1in); length 15.53m (50ft 11in); height 4.80m (15ft 9in); wing area 33.7m2 (362.75 sq ft)
Armament:	one 30mm (1.18in) GSh-30-2 cannon; eight external pylons with provision for up to 4400kg (9700lb) of stores, plus two outboard pylons for AAMs

Mikoyan MiG-29 Fulcrum

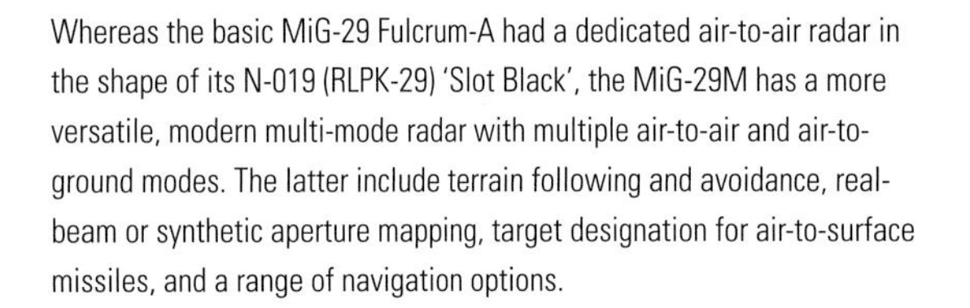

Whereas the basic MiG-29 Fulcrum-A had a dedicated air-to-air radar in the shape of its N-019 (RLPK-29) 'Slot Black', the MiG-29M has a more versatile, modern multi-mode radar with multiple air-to-air and air-to-ground modes. The latter include terrain following and avoidance, real-beam or synthetic aperture mapping, target designation for air-to-surface missiles, and a range of navigation options.

The MiG-29 has been criticized because it has a comparatively old-fashioned cockpit, with conventional analogue instruments, and without any multi-fuction display screens. Many pilots, however, feel that more modern cockpits tend to over-saturate the pilot with information.

A wide variety of ordnance has been displayed on the MiG-29. For interception and air superiority missions, there is normally a mix of AA-9 long-range 'fire and forget' missiles, which are similar to the US Navy's Phoenix, and AA-10 Alamo medium-range missiles. The missile being launched here is an AA-8 Aphid short-range AAM.

Forty per cent of the MiG-29's lift is provided by its lift-generating centre fuselage and the aircraft is able to achieve angles of attack at least 70 per cent higher than earlier fighters.

Above: **The Federal German Luftwaffe inherited a number of MiG-29s from the former German Democratic Republic, and the best of these aircraft – two of which are pictured here – were assigned to 731 Staffel of *Jagdgeschwader* (JG) 73 at Laage.**

The appearance in the early 1980s of the MiG-29, with its superb agility and its apparent ability to perform combat manoeuvres that could not be matched by any aircraft in the West, came as an unpleasant surprise to NATO. Just as the F-15 was developed to counter the MiG-25 Foxbat and the MiG-23 Flogger, both of which were unveiled in the late 1960s, the MiG-29 Fulcrum and another Russian fighter, the Sukhoi Su-27 Flanker, were designed in response to the F-15 and its naval counterpart, the Grumman F-14 Tomcat. Both Russian aircraft share a similar configuration, combining a wing swept at 40 degrees with highly swept wing root extensions, underslung engines with wedge intakes, and twin fins. MiG-29 design emphasis from the start was on very high manoeuvrability and the ability to destroy targets at distances of between 200m (660ft) and 60km (32nm). The aircraft has an RP-29 pulse-Doppler radar capable of detecting targets at around 100km (62 miles) against a background of ground clutter. Fire control and mission computers link the radar with a laser rangefinder and infrared search/track sensor, in conjunction with a helmet-mounted target designator. The radar can track ten targets simultaneously, and the system allows the MiG-29 to approach and engage targets without

emitting detectable radar or radio signals. The Fulcrum-A became operational in 1985. The MiG-29K is a navalized version, the MiG-29M is a variant with advanced fly-by-wire systems, and the MiG-29UB is a two-seat operational trainer. In 2004 India ordered 12 MiG-29K single-seat and four MiG-29KUB two-seat fighters. In January 2010, India and Russia signed a deal worth US$1.2 billion for the Indian Navy to receive an additional 29 MiG-29Ks.

The Russian Air Force has begun an upgrade programme for 150 of its MiG-29 fighters, which will be designated MiG-29SMT. The upgrade comprises increased range and payload, a new glass cockpit, new avionics, improved radar, and an in-flight refuelling probe. The radar will be the Phazotron Zhuk (Beetle); it will also be able to track ten targets simultaneously, but out to a range of 245km (152 miles). A two-seater version, the MiG-29M2, has also been demonstrated, as has a super-manoeuvrable variant, the MiG-29OVT, with three-dimensional thrust-vectoring engine nozzles. Twenty-two MiG-29s of the Polish Air Force are also to be upgraded by EADS (formerly Daimler Chrysler Aerospace). Modifications are necessary to bring the aircraft up to NATO standards. EADS has carried out similar modifications to MiG-29s of the former East German Air Force, now in Luftwaffe service, and has joined with RAC-MiG to offer modernization packages to other MiG-29 customers.

The MiG-29 is equipped with two RD-33 turbofan engines, and is the first aircraft in the world to be fitted with dual-mode air intakes. During flight, the open air intakes feed air to the engines in the normal way, but while

Below: **The MiG-29M seen here has an increased-volume fuselage spine containing extra fuel and avionics, displaced by other internal changes. It also has an increased-area tailplane, giving greater control authority in pitch and roll.**

Right: **Another view of a MiG-29 of the German Luftwaffe. German MiG-29s have been progressively upgraded since 1999, the aircraft emerging as MiG-29Gs. Twelve single-seaters and a pair of MiG-29UB two-seat trainers were in service in 2002.**

the aircraft is taxiing the air intakes are closed and air is fed through the louvres on the upper surface of the wing root to prevent ingestion of foreign objects from the runway, which is particularly important when operating from unprepared airstrips.

About 600 MiG-29s are in service with the Russian Air Force, and the type also serves with the air forces of Bangladesh (8), Belarus (50), Bulgaria (17), Cuba (18), Eritrea (5), Germany (19), Hungary (21), India (70), Iran (35), Kazakhstan (40), Malaysia (16), Myanmar (10), North Korea (35), Peru (18), Poland (22), Romania (15), Slovakia (22), Syria (50), Turkmenistan (20), Ukraine (220), Uzbekistan (30) and Yemen (24).

After the break-up of the Soviet Union, one of the smallest nations to inherit the MiG-29 was Moldova, a state sandwiched between the Ukraine on the one hand and Romania on the other, which did not have the resources to operate the type successfully. Twenty-one Fulcrums were purchased from Moldova by the USA in 1997, partly as a measure to prevent them from falling into Iranian hands, as 14 of the aircraft were Fulcrum-Cs, which have a tactical nuclear capability. The aircraft were dismantled and shipped in pairs in the capacious cargo holds of C-17 Globemaster III aircraft, which flew the Fulcrums to the National Air Intelligence Center at Wright-Patterson ASFB, Dayton, Ohio.

Specification: Mikoyan-Gurevich MiG-29M	
Type:	single-seat air superiority fighter
Powerplant:	two 9409kg (20,725lb) thrust Sarkisov RD-33K turbofans
Performance:	maximum speed 2300km/h (1430mph) at 11,000m (36,090ft); service ceiling 17,000m (55,775ft); range 1500km (932 miles) on internal fuel
Weights:	empty 10,900kg (24,030lb); maximum take-off 18,500kg (40,792lb)
Dimensions:	wing span 11.36m (37ft 3in); length 17.32m (56ft 10in); height 7.78m (25ft 6in)
Armament:	one 30mm (1.18in) GSh-30 cannon; eight external hardpoints with provision for up to 4500kg (9922lb) of stores, including six AAMs, rocket pods, bombs etc

Sukhoi Su-27 Flanker

The Su-27 is usually armed with long-range R-27 air-to-air missiles on the inner hardpoints and fuselage stations, and short-range R-73s on the wingtip rails and outer wing hardpoints.

This Sukhoi Su-27 Flanker-B is unusual in that it displays a large shark design along the port engine bay, which may have been applied for the aircraft's participation in the massive 1995 VE-Day fly-past over Moscow. The aircraft is operated by the 760th Air Fighting Development Regiment, which is primarily entrusted with the training of weapons instructors and the development of tactics.

The Flanker-B is equipped with a sophisticated weapons control system using an RLPK-27 coherent pulse-Doppler jam-proof radar with track-while-scan and look-down/shoot-down capability. In case of radar failure, the pilot is backed up by an electro-optical system containing a laser rangefinder and infrared search and track system, which can be attached to his helmet-mounted target designator.

Pre-series Flanker-Bs sported a square-cut flat top to the fins, with a large anti-flutter mass protruding forward. Early series production Flanker-Bs retained the anti-flutter mass, but displayed pointed fins. This aircraft is a late series Flanker-B, which has 'pointed' fins with no anti-flutter masses.

Together with the MiG-29, the Sukhoi Su-27, code-named Flanker under the NATO reporting system, has provided the Soviet Union/CIS with a formidable air defence capability since the mid-1980s. The Su-27, like the F-15, is a dual-role aircraft; in addition to its primary air superiority task it was designed to escort Su-24 Fencer strike aircraft on deep penetration missions. The project originated in 1969, the first prototype, designated T10-1, flying on 20 May 1977. The T10-1 did not meet its requirements and had numerous problems, and the second prototype, the T10-2, crashed owing to a systems failure, killing its pilot. After much redesign a new prototype, the T10-S, made its appearance, flying for the first time on 20 April 1981. It was this aircraft that evolved into the Su-27. Full-scale production of the Su-27P Flanker-B air defence fighter began in 1980 on a limited scale and the type entered full production in 1982, but the aircraft did not become fully operational until 1984. Today, the type is in service with Russia, Ukraine, Belarus, Kazakhstan and Vietnam, and is built under licence in China as the F-11. A variant, the Su-30MK, has been sold to India with licensed local production; 50 aircraft were ordered, the first due to be delivered in July 2002. China has also ordered 30 examples of the two-seat Su-30MKK.

The Su-27 is a highly integrated twin-finned aircraft. Its airframe is constructed of titanium and high-strength aluminium alloys. The engine nacelles are fitted with trouser fairings to provide a continuous streamlined profile between the nacelles and the tail beams. The central beam section between the engine nacelles consists of the equipment compartment, fuel tank and the brake parachute container. The forward fuselage is of semi-monocoque construction and includes the cockpit, radio compartments and avionics bay. Like its contemporary, the MiG-29 Fulcrum, the Su-27 combines a wing swept at 40 degrees with highly swept wing root extensions, underslung engines with wedge intakes, and twin fins. The combination of modest wing sweep with highly swept root extensions is designed to enhance manoeuvrability and generate lift, making it possible to achieve quite extraordinary angles of attack. The Su-27UB Flanker-C is a two-seat training version. The Sukhoi Su-35, derived from the Flanker-B and originally designated Su-27M, is a second-generation version with improved agility and enhanced operational capability.

Left: **This Su-27 Flanker bears the distinctive paint scheme of the Russian Knights, the Russian Air Force's display team. The Su-27's thrust-vectoring nozzles allow it to demonstrate an extraordinary range of manoeuvres, making it an ideal display aircraft.**

Above: **The Su-27K, pictured here, features new double-slotted flaps which span almost the entire length of the trailing edge. The inboard flap sections operate differentially as drooping ailerons at low speeds. The Su-27K is optimized for carrier operations.**

The latest version of the Flanker is the upgraded Su-27SK, which is equipped with a new electronic countermeasures suite for individual aircraft, and for mutual and group protection in the forward and rear hemispheres. The countermeasures system includes a pilot illumination radar warning receiver, chaff and infrared decoy dispensers, and an active multi-mode jammer located in the wingtip pods. The aircraft is fitted with a Phazotron N001 Zhuk coherent pulse-Doppler radar with track-while scan and look-down/shoot-down capability. The range of the radar against 3-square-metre targets is over 100km (62 miles) in the forward hemisphere and 40km (25 miles) in the rear hemisphere. The radar has the capacity to search, detect and track up to ten targets with automatic threat assessment and prioritization. The aircraft has an OEPS electro-optical system, which includes an infrared search-and-track (IRST) sensor collimated with a laser rangefinder. The range of the electro-optical system is 40 to 100km, depending on the aspect angle presented by the target. The radio communications suite provides voice and data; VHF/UHF radio communications between aircraft and ground control stations within sight range; voice radio communication with ground control stations and between aircraft up to a range of 1500km (930 miles); an encrypted data link for the exchange of combat information between aircraft; and command guidance from ground control stations using automatic interception mode.

The Su-27SK is powered by two AL-31F turbofan engines, designed by the Lyulka Engine Design Bureau (NPO Satum). Each engine has two air intakes: a primary wedge intake and a louvred auxiliary air intake, the latter being used when the aircraft is taxiing to avoid the danger of ingesting foreign objects. The twin-shaft, turbofan engine has after-turbine flow mixing, a common afterburner, an all-mode variable area jet exhaust nozzle, an independent start and a main electronic control, and a reserve hydromechanical engine mode control system. The high-temperature sections of the engine are made of titanium alloy.

The aircraft now known as the Sukhoi Su-35, derived from the Flanker-B and originally designated Su-27M, is a second-generation version with improved agility and enhanced operational capability.

Opposite: **The basic Su-27 design has produced several advanced derivatives, of which the most fundamentally changed is the Su-27IB (Su-34) long-range strike aircraft. This combines the Su-27s fuselage with a new forward section featuring side-by-side seating and a distinctive 'platypus' nose.**

Specification: Sukhoi Su-27 Flanker-B	
Type:	single-seat air superiority fighter
Powerplant:	two 12,500kg (27,562lb) thrust Lyulka AL-31F turbofans
Performance:	maximum speed 2500km/h (1552mph) at high altitude; service ceiling 18,000m (59,055ft); range 4000km (2485 miles)
Weights:	empty 20,748kg (45,750lb); maximum take-off 30,000kg (66,150lb)
Dimensions:	wing span 14.70m (48ft 2in); length 21.94m (71ft 11in); height 6.36m (20ft 10in); wing area 46.5m2 (500 sq ft)
Armament:	one 30mm (1.18in) GSh-301 cannon; ten external hardpoints with provision for various combinations of AAMs

Index

Afghanistan 35, 74, 92, 241
Angola 157
Argentina 153, 160–1, 160–1
  Falklands War (1982) 116–17, 119, 141, 153, 160–1
Australia 65, 83, 115, 152, 153
Austria 203
Avro (Hawker Siddeley) Vulcan 116–19
  B.Mk.1 118
  B.Mk.1A 118–19
  B.Mk.2 118–19
  B.2 XM597 116–17
  B.2 XM605 118
  B.Mk.2A 119
  VX770 prototype 118
  VX777 prototype 118
Balkans 74, 78, 92
Bennett, E.N.K. 126
Boeing B-52 Stratofortress 32–5
  B-52A 34
  B-52B 34
  B-52C 34
  B-52D 34, 35, 35
  B-52E 34
  B-52F 34
  B-52G 34–5, 34
  B-52H 32–3, 35
  XB-52 prototype 34
  YB-52 prototype 34
Brazilian Air Force 52–3
British Aerospace Harrier, T.Mk.52 136
British Aerospace (Hawker Siddeley) Hawk 134–7
  demonstrator 136
  Hawk 200 136
  Hawk Series 60 136
  Hawk Series 100 136
  T.Mk.1 136
  T.Mk.1A 134–5, 136, 137
British Aerospace Sea Harrier 138–41
  FA.2 140–1, 141
  FRS.1 138–41
  XZ438 140
  XZ439 140
  XZ440 140
  XZ450 140
  XZ451 140
Brown, 1st Lt Russell 15
Burbank factory (Skunk Works) 30, 51

Canada/Canadair 23, 27, 39, 55, 83
Carter, Jimmy 96
Charyk, Dr Joe 30
Chasan dam, North Korea 19
China 31, 218, 219, 251
CIA 30–1
Cuban missile crisis (1962) 31, 219
Czechoslovak Air Force 232–3

Danish Air Force 200–1, 203
Dassault/Dornier Alpha Jet 178–81
  Alpha Jet 2 181
  Alpha Jet A 178–9, 180
  Alpha Jet E 180, 180
  Alpha Jet NGEA 181
Dassault, Marcel 143
Dassault Mirage 2000C 162–5
  Mirage 2000B 164–5, 164–5
  Mirage 2000C-1 165
  Mirage 2000H 162–3, 165
  Mirage 2000N 142–3, 165, 165
Dassault Mirage F.1 154–7
  Mirage F.1A 156
  Mirage F.1AZ 154–5, 156, 156, 157
  Mirage F.1B 156
  Mirage F.1C 156
  Mirage F.1CT 156–7
  Mirage F.1CZ 157
  Mirage F.1EQ5 157
Dassault Mirage III 148–53
  Mirage I 150
  Mirage II 150
  Mirage IIIA 150–1
  Mirage IIIB 151
  Mirage IIIBZ 152
  Mirage IIIC 150, 151
  Mirage IIICJ 152
  Mirage IIICZ 148–9, 152
  Mirage IIID 152, 153
  Mirage IIID2Z 152
  Mirage IIIDZ 152
  Mirage IIIE 150–1, 152–3
  Mirage IIIEE 152–3, 153
  Mirage IIIEZ 152
  Mirage IIIO 153
  Mirage IIIP 153
  Mirage IIIR 153
  Mirage IIIRZ 152
  Mirage IIIS 153
  Mirage G 153
  Mirage G8 153
Dassault Mystère IVA 144–7
  Mystère IVB 147
  MD.452 Mystère IIC 146
  Super Mystère B.2 147
Dassault Rafale 166–71
  Rafale-A 169, 169
  Rafale-B 169
  Rafale B.301 168
  Rafale-C 169, 171
  Rafale-M 169, 170–1
Dassault Super Etendard 158–61
  Etendard IVM-01 160
  Etendard IVP 160
De Havilland Vampire 112–15
  DH.100 114
  F.1 114
  F.3 114–15
  F.30/31 115
  FB.5 115
  FB.6 114–15
  FB.9 112–13, 115
  FB.31 115
  FB.52 115
  F.Mk.4 115
  NF.10 115
  Sea Vampire F.20 & F.21 115
  T.11 114, 115
  T.Mk.55 114
Duke, Neville 123
Dulles, Allan 30

Edens, Lt Bruce 233
Egypt 115, 146, 147, 181, 219, 227
English Electric Lightning 128–33
  F.Mk.1 131
  F.Mk.3 128–9, 133
  F.Mk.6 130, 131, 132–3
  P.1A 131
  P.1B 131
  T.Mk.5 130, 131
Eurofighter Typhoon 8, 192–7
  DA.1 172–3, 196–7
  DA.2 194–5
Experimental Aircraft Programme (EAP) 194–5

Fairchild Republic A-10 Thunderbolt II 9, 66–9
  A-10A 68, 69
Falklands War (1982) 116–17, 119, 141, 153, 160–1
Finland 115, 203
Ford, Gerald 96
France
  aircraft of 8, 142–71
  international collaborations 174–81, 192–7
Freckleton, Lt William C. 44–5

General Dynamics F-111 62–5
  F-111A 64–5
  F-111B 64
  F-111C 65, 65
  F-111D 65
  F-111E 65
  F-111F 62–3, 65
Germany 38,
  aircraft of 9
  international collaborations 178–85, 192–7
  jet engine technology development 7–8
  use of Soviet aircraft 244, 246–7
Grumman A-6 Intruder 40–3

A-6A prototype 42
A-6A 42
A-6C 43
A-6E 40–1, 43
EA-6A 42–3
EA-6B Prowler 42–3
KA-6D tanker 43
Grumman F-14A Tomcat 70–5
F-14B 74
F-14D 72–3, 74, 75
Gulf War (1991) 35, 43, 69, 74, 78, 86, 87, 92, 100, 127, 157, 174–5, 237

Handley Page Victor 124–7
B.2(BS) 126
B.(K).Mk 1 127
B.(K).Mk 1A 127
B.Mk.1 127
B.Mk.1A 127
B.Mk.2 127
B.(PR).Mk.1 127
B.(PR).Mk.2 127
HP.80 Victor 126–7
K.Mk.2 127
Hartley, Keith 172–3, 196–7
Have Blue project 90
Hawker Hunter 120–3
F.4 123
FGA.9 122, 123
F.Mk.1 122–3
F.Mk.2 123
F.Mk.5 123
F.Mk.6 123
F.Mk.7 123
F.Mk.8 123
F.F.Mk.12 123
F.Mk.73 120–1
FR.10 123
GA.11 123
P.1067 122, 123
P.1083 123
T.Mk.8 123
T52 123
T62 123
T66 123
T67 123
T69 123
Hazelden, Sqn Ldr H.G. 126
Hungary 219

Ilyushin Il-28 Beagle 216–19
Il-28R 219
Il-28T 219
Il-28U 219
India 115, 123, 147, 162–3, 165, 177, 228–9, 245, 246, 251
wars with Pakistan 38, 123, 147, 153
Iran 47, 54–5, 157, 160–1, 246
Iraq 47, 157, 160–1, 237, 240
Gulf War (1991) 35, 43, 69, 74, 78, 86, 87, 92, 100, 127, 157, 174–5, 237
Operation Desert Storm (1991) 58, 61, 65, 93, 174–5, 177
Israel 47, 78, 86, 146, 147, 151–2, 227
Italy 38–9, 115
international collaborations 182–5, 192–7

Jabara, Lt Col James 23
Japan 36–7, 39, 47, 86, 100
Johnson, Clarence 'Kelly' 30, 51
Johnson, Lyndon B. 50

Korean War (1950–53) 15, 16–17, 19, 23, 230

Larson, Maj George W. 96
Lebanon 78, 161
Leenhouts, Commander John 58
Libya 42, 51, 65, 74, 80–1, 83, 153
Lockheed F-80 Shooting Star (formerly P-80) 12–15
F-80A 14–15
F-80B 14, 15
F-80C 12–13, 15
RF-80A 15
RF-80C 15
Lockheed F-104 Starfighter 36–9
F-104A 38
F-104B 38
F-104C 38–9
F-104D 39
F-104F 39
F-104G 38, 39
F-104J 36–7, 39
F-104S 38–9, 39
XF-104 prototypes 38
YF-104 prototypes 38
Lockheed F-177A Night Hawk 88–93
Lockheed-Martin F-16 76–9
F-16B 78
F-16C 76–7, 78, 79
F-16D 78
F-16I 78
F-16N 79
Sniper XP 79
Lockheed Martin F-22 Raptor 9, 102–5
F-22A 104–5, 104
Lockheed Martin F-35 Lightning II 106–9
F-35A 108, 109
F-35B 108, 108, 109
F-35C 108, 109
Lockheed SR-71A Blackbird 48–51
A-12 51
SR-71B 50, 51
YF-12A 51
Lockheed U-2 28–31
U-2R 28–9, 30–1

McDonnell Douglas/British Aerospace Harrier II 186–91
AV-8A 189, 190
AV-8B 188, 190, 190–1
GR.Mk.1 189
GR.Mk.7 186–7, 190
Harrier GR5 190
P.1127 189
YAV-8B Harrier II 189
McDonnell Douglas F-4 Phantom II 44–7, 230, 233
F-4B 44–5, 46
F-4C 46
F-4D 46
F-4E 46, 47
F-4F 46–7
F-4G Wild Weasel 46–7, 47
F-4J 47
F-4K 47
F-4M 47
RF-4B 46
RF-4C 46
McDonnell Douglas F-15 Eagle 9, 84–7
F-15A 86
F-15B 86
F-15C 86–7
F-15E Strike Eagle 84–5, 86, 86, 87
F-15I 86
F-15J 86
F-15S 86
McDonnell Douglas F/A-18 Hornet 8, 80–3
EA-18G Growler 83
F/A-18A 80–1, 83
F/A-18B 83
F/A-18C 82, 83
F/A-18D 82, 83
F/A-18E/F Super Hornet 83
McDonnell F-101 Voodoo 24–7
F-101A 27
F-101B 24, 26, 27, 27
F-101C 27
XF-88 prototype 26–7
McDougall, Squadron Leader 116–17
Messerschmitt Me 262 6–7, 8
Mikoyan-Gurevich MiG-17 Fresco 220–3
17F (Fresco-C) 222, 223
17P (Fresco-B) 222
17PFU 222
Fresco-A 222
Mikoyan-Gurevich MiG-21 Fishbed 228–33
21B (Fishbed-L) 230
21F 230
21M 231
21MF 228–9
21MF Fishbed-J 230–1, 233
21U 230
Fishbed-A 230
Fishbed-B 230
Fishbed-C 230
Fishbed-N 230
Mikoyan MiG-15 8, 15, 19, 23
Mikoyan MiG-25 Foxbat 86, 234–7
25BM 237
25P (Foxbat-A) 236–7
25R 237
25RB 237
Mikoyan MiG-29 Fulcrum 242–7
29G 246–7
29K 245
29M 245, 245, 246
29M2 245
29OVT 245
29SMT 245
29UB 245
Fulcrum-A 245

Fulcrum-C 246
Mikoyan MiG-31 (Foxhound) 237
Mixed Fighter Force 136–7
Moldova 246
Murphy, Cap Anthony R. 87

New Zealand 115
Nigeria 177, 219, 222
night attack capabilities 83, 190, 230
North American F-86 Sabre 20–3
F-86A 23, 23
F-86C 23
F-86D 20–1, 23
F-86E 23
F-86F 23
F-86H 23
F-86J 23
F-86K 23
F-86L 23
XP-86 Sabre 22–3, 22
North American NA-134 22
North American NA-140 22
North Korea 15, 19
see also Korean War
North Vietnam 31, 35, 219, 223
see also Vietnam War
Northrop F-5 Tiger 52–5
F-5A Freedom Fighter 54
F-5A Tiger 55
F-5E Tiger II 52–3, 55
F-5E Tiger IV 55
F-20 Tigershark 55
RF-5E TigerEye 55
T-38A Talon 54
YF-23 55
Northrop Grumman B-2 Spirit 98–101
Norway 55, 115

Oman 120–1, 177
Operation Desert Storm (1991) 58, 61, 65, 93, 174–5, 177

Pakistan 38, 123, 147, 153, 241
Panavia Tornado ADV 182–5
Tornado F.2 184
Tornado F.3 182–3, 184–5
Parsons, Col Rick 87
Poland 227, 227, 246
Powers, Francis G. 31

Reagan, Ronald 96
reconnaissance aircraft 28–31, 48–51, 207
Red Arrows 137
Republic F-84 Thunderjet 16–19
F-84B 18
F-84C 18
F-84D 19
F-84E 19
F-84G 16–17, 19, 19
Rockwell B-1B Lancer 9, 10–11, 94–7
B-1A 96, 97
Romania 231
Rumsfeld, Donald H. 96
Russia *see* USSR/Russia
Russian Knights 250–1
Rutskoi, Col Alexander V. 241

Saab A-21R 202
Saab A-29 203
Saab A-32 Lansen 203
Saab J-21R 202
Saab J-29 199, 202–3
Saab J-32B 203
Saab J-35 Draken 199, 200–3
J-35A 203
J-35C 203
J-35D 203
J-35F 203
J-35J 203, 203
R-35 200–1
RF-35 203
Saab JA-37 Viggen 204–7
AJ-37 206
SF-37 207
SH-37 207
SK-37 207
Saab JAS-39A Gripen 198–9, 208–13
37-51 210
39-1 210
39-2 210
39-3 210
39-4 210
39-5 210
39-102 210
39B 210
Saab S-29C 203
Saudi Arabia 86, 131
Schillereff, Capt Raymond 15
Schilling, Lt Col David 14
SEPECAT Jaguar 174–7
Jaguar A 176
Jaguar B 176
Jaguar E 176
Jaguar GR.Mk.1 176, 176
Jaguar GR.Mk.1A 177
Jaguar International 177
Jaguar S 176
Jaguar T.Mk.2 176
short take-off, vertical landing (STOVL) 189
Simpson, Lt Jim 16–17
Six Day War (1967) 147, 227
South Africa 115, 148–9, 152, 153, 154–5, 157, 213
Soviet Union *see* USSR/Russia
Spain 83, 153, 190, 192–7
Suez campaign (1956) 147, 147, 219
Sukhoi, Pavel 226
Sukhoi Su-7 Fitter 224–7
AF Su-7BM 227
Su-7B Fitter-A 226, 227
Su-17/20 Fitter C 226
Su-17M 226
Su-22M 227
Sukhoi Su-9 Fishpot-A 226, 227
Sukhoi Su-11 Fishpot-B 227
Sukhoi Su-11 Fishpot-C 227
Sukhoi Su-25 Frogfoot 238–41
Frogfoot A 241
Su-25K 241
Su-25TM 238–9, 240–1, 241
Su-25UBK 241
Su-25UBT 241
Sukhoi Su-27 Flanker 248–53
F-11 251
Flanker-B 248–9, 251, 253
Su-27IB 252
Su-27K 253
Su-27M 214–15
Su-27SK 253
Su-27UB Flanker-C 251
Su-30MK 251
Su-30MKK 251
Su-35 251, 253
T10-1 251
T10-2 251
T10-S 251
Sweden
aircraft of 198–213
Switzerland 115
Syria 86, 105, 222, 236

Toksan dam, North Korea 19
Tucker, Capt. R.E. 230, 233
Tupolev Tu-85 Bear 132–3

United Kingdom
aircraft of 7–9, 111–41
buying aircraft overseas 46, 47
international collaborations 174–7, 182–97
United States
aircraft of 8, 9, 10–109
international collaborations 186–91
USSR/Russia
aircraft of 8, 9, 214–53

Vietnam War (1955–75)
American aircraft 31, 35, 42, 44–5, 51, 58–9, 59, 64–5
North Vietnamese aircraft 219, 223, 230, 233
Vought A-7 Corsair II 56–61
A-7A 59
A-7B 59
A-7D 56–7, 59, 60–1, 61
A-7E 58, 59, 61
A-7K 59, 61
Vulcan B.2 XM657 110–11

War of Attrition (1969–70) 227
Warsitz, Erich 7
Weigand, Lt Garry 44–5
World Absolute Air Speed Record 123

Yom Kippur War (1973) 47, 227